T0222876

Machine Learning in the Oil and Gas Industry

Including Geosciences, Reservoir Engineering, and Production Engineering with Python

Yogendra Narayan Pandey
Ayush Rastogi
Sribharath Kainkaryam
Srimoyee Bhattacharya
Luigi Saputelli

Apress®

Machine Learning in the Oil and Gas Industry

Yogendra Narayan Pandey
Houston, TX, USA

Ayush Rastogi
Denver, CO, USA

Sribharath Kainkaryam
Houston, TX, USA

Srimoyee Bhattacharya
Houston, TX, USA

Luigi Saputelli
Houston, TX, USA

ISBN-13 (pbk): 978-1-4842-6093-7
https://doi.org/10.1007/978-1-4842-6094-4

ISBN-13 (electronic): 978-1-4842-6094-4

Managing Director, Apress Media LLC: Welmoed Spahr
Acquisitions Editor: Celestin Suresh John
Development Editor: James Markham
Coordinating Editor: Aditee Mirashi

Cover designed by eStudioCalamar

Cover image designed by Freepik (www.freepik.com)

Distributed to the book trade worldwide by Springer Science+Business Media New York, 1 New York Plaza, Suite 4600, New York, NY 10004-1562, USA. Phone 1-800-SPRINGER, fax (201) 348-4505, e-mail orders-ny@springer-sbm.com, or visit www.springeronline.com. Apress Media, LLC is a California LLC and the sole member (owner) is Springer Science + Business Media Finance Inc (SSBM Finance Inc). SSBM Finance Inc is a **Delaware** corporation.

For information on translations, please e-mail booktranslations@springernature.com; for reprint, paperback, or audio rights, please e-mail bookpermissions@springernature.com.

Apress titles may be purchased in bulk for academic, corporate, or promotional use. eBook versions and licenses are also available for most titles. For more information, reference our Print and eBook Bulk Sales web page at http://www.apress.com/bulk-sales.

Any source code or other supplementary material referenced by the author in this book is available to readers on GitHub via the book's product page, located at www.apress.com/978-1-4842-6093-7. For more detailed information, please visit http://www.apress.com/source-code.

Printed on acid-free paper

To my family.

—Yogendra

To my family for their patience and support through the journey of writing this book.

—Ayush

To my family.

—Sribharath

For my family, who provides me the energy and inspiration all the time.

—Luigi

To my family, for all their love and support.

—Srimoyee

To all geoscientists and petroleum engineers who wish to have a head start on machine learning problems and to the younger generation of data scientists who need a quick reference on oil and gas industry use cases.

—The authors

Table of Contents

About the Authors

Yogendra Narayan Pandey is a senior product manager at Oracle Cloud Infrastructure. Previously, he founded an oil and gas machine learning consultancy, Prabuddha LLC. He has more than 15 years of experience in orchestrating intelligent solutions for the oil and gas, utilities, and chemical industries. He has worked in different capacities with oil and gas and utility companies, including Halliburton, ExxonMobil, and ADNOC. Yogendra holds a bachelor's degree in chemical engineering from the Indian Institute of Technology (BHU) and a PhD from the University of Houston, specializing in high-performance computing applications for complex engineering problems. He has served as an executive editor for the *Journal of Natural Gas Science and Engineering*. Also, he has authored/co-authored more than 25 peer-reviewed journal articles, conference publications, and patent applications. He is a member of the Society of Petroleum Engineers.

Ayush Rastogi is a data scientist at BPX Energy (BP Lower48) based in Denver, CO. His research experience is in the area of mechanistic modeling for multiphase fluid flow while integrating physics-based and data-driven algorithms to develop robust predictive models. Some of his other areas of interest include gas well deliquification, liquid loading, mechanistic models, hydraulic fracturing, production data analytics, and machine learning applications in the oil and gas industry. His prior work experience is at Liberty Oilfield Services, where he was a technology and field engineer intern in Texas, North Dakota, and Colorado. He also has experience working as a petroleum engineering consultant in Houston, TX. He has published several papers for the Society of Petroleum Engineers and the *Journal of Natural Gas Science and Engineering*. Ayush holds a BS in petroleum engineering from Pandit Deendayal Petroleum University, India (2012), MS in petroleum engineering from the University of Houston (2014), and a PhD in petroleum engineering from Colorado School of Mines (2019) with a minor in Computer Science, focusing on Data Analytics and Statistical Modeling. He is an active member of the Society of Petroleum Engineers. Ayush can be reached at ayush.rastogi@outlook.com.

 **Sribharath Kainkaryam** is a data scientist at TGS in Houston, TX. His interests are in the application of computer vision techniques on seismic imaging problems. His research experience is in the area of seismic imaging, anisotropy, and high-performance computing. He was a research geophysicist at Schlumberger, working on problems in imaging, velocity model building, and the development of data-driven algorithms for modeling wave propagation in subsurfaces. He graduated with an MS in computational geophysics from Purdue University. Prior to his graduate degree, he graduated from the Indian Institute of Technology, Kharagpur. He has been a reviewer for *Geophysics*, *Geophysical Prospecting*, *The Leading Edge*, and several other journals. He is a member of the Society of Exploration Geophysicists and the European Association of Geoscientists and Engineers.

Luigi Saputelli is a reservoir engineering expert advisor for the Abu Dhabi National Oil Company (ADNOC). He has 30 years of experience. He has held various positions as a reservoir engineer (integrated reservoir management, simulation, improved oil recovery projects, field development), drilling engineer (drilling and well planning projects, drilling rig automation), and production engineer (production modeling, engineering, and operations workflow automation projects) at various operators and services companies around the world, including ADNOC, PDVSA, Hess, and Halliburton. Saputelli is an industry-recognized researcher, invited lecturer, and active volunteer. He has served on the Society of Petroleum Engineers (SPE) *Journal of Petroleum Technology* (JPT) editorial committee as the data communication and management technology feature editor since 2012 and on the SPE Production and Operations Advisory Board since 2010. He is a founding member of the SPE Real-Time Optimization Technical Interest Group and the Petroleum Data-Driven Analytics Technical section. He is the recipient of the 2015 SPE International Production and Operations Award. He has published more than 100 industry papers on applied technologies related to digital oilfields, reservoir management, real-time optimization, and production operations. Saputelli holds a BSc in electronic engineering from Universidad Simon Bolivar (1990), an MSc in petroleum engineering from Imperial College (1996), and a PhD in chemical engineering from the University of Houston (2003). He serves as a managing partner at Frontender, a petroleum engineering services firm based in Houston. He can be reached at LSaputelli@frontender.com.

Srimoyee Bhattacharya is a reservoir engineer for the Permian Asset team at the Shell Exploration & Production Company. She has more than 12 years of combined academic and professional experience in the oil and gas industry. She has experience in field development planning, production surveillance, decline curve analysis, reserves estimation, reservoir modeling, enhanced oil recovery, history matching, fracture design, production optimization, proxy modeling, and applications of multivariate analysis methods. She also worked with Halliburton as a research intern on the digitalization of oil fields and field-wide data analysis using statistical methods. Srimoyee holds a PhD in chemical engineering from the Cullen College of Engineering at the University of Houston and a bachelor of technology in chemical engineering from the Indian Institute of Technology, Kharagpur, India. She served as a technical reviewer for the *SPE Journal*, the *Journal of Natural Gas Science and Engineering*, and the *Journal of Sustainable Energy Engineering*. She has authored/co-authored more than 25 peer-reviewed journal articles, conference publications, and patent applications.

About the Technical Reviewer

Sukanya is a data scientist working for Capgemini. She has extensive experience working with the Internet of Things (IoT), building various solutions. She enjoys working on the intersection of IoT and data science. She leads the PyData Mumbai (an international educational program focused on an open data science ecosystem) chapter, PyLadies Mumbai (an international mentorship group for women in Python—supported by the Python Software Foundation) chapter, and the AWS User Group, Mumbai. Besides work and community efforts, Sukanya loves to explore new tech and pursue research. She has published white papers on IEEE. She can be reached on LinkedIn at `www.linkedin.com/in/sukanyamandal`.

Introduction

This book is for anyone interested in applications of machine learning in the oil and gas industry. The contents of the book are organized to provide foundational knowledge of computer programming, the implementation of machine learning and deep learning algorithms, and the ways and means to use them for solving oil and gas industry problems. This book does not provide production-quality codes, but it equips you with information on how you can start building a machine learning solution for the oil and gas industry problems discussed in this book and beyond. This book is organized into eight chapters, as follows.

Chapter 1 overviews the different segments of the oil and gas industry and provides example applications of machine learning to upstream industry problems.

Chapter 2 is a brief primer on the Python programming language. You become familiar with the basic syntaxes of the Python programming language. Afterward, you learn about Python libraries, such as NumPy, pandas, and Matplotlib, which are useful for certain machine learning–related tasks.

Chapter 3 overviews machine learning and deep learning concepts. In this chapter, you learn about supervised and unsupervised machine learning concepts. You work on examples using simplistic and clean datasets. The scikit-learn and TensorFlow libraries are used for implementing machine learning and deep learning code samples, respectively.

Chapter 4 focuses on the application of machine learning to problems in seismic processing and interpretation. Using seismic data available from open data sources, we implement automated salt interpretation. Without going into implementation details, strategies for solving other seismic interpretation problems are briefly discussed.

Chapter 5 focuses on the problems related to geological modeling, including supervised machine learning for estimation of the petrophysical properties away from well locations.

Chapter 6 focuses on approaches for developing machine learning models in decline curve analysis and the use of these models in production forecasting.

Chapter 7 covers production modeling using machine learning methodologies, including production optimization. Part of the chapter provides a methodology for virtual metering and the formulation of virtual sensors.

Chapter 8 discusses opportunities, challenges, and expected future trends. In this chapter, we discuss the challenges encountered when executing machine learning–based digital transformation projects in the oil and gas industry. The potential pitfalls that lead to project failure and ways to avoid them are discussed. Also, the opportunities that inherently lie in addressing these challenges are discussed from both an executive's and a practitioner's perspective. A roadmap of the machine learning–enabled digital transformation of the upstream industry over the coming years is provided.

CHAPTER 1

Toward Oil and Gas 4.0

Energy is one of the fundamental needs for sustaining modern human life. In the year 2016, global energy demand was 552 quadrillion British thermal units (quads). Out of this, 177 quads were supplied by oil, and 127 quads were supplied by gas. This indicates that 55% of global energy demand in 2016 was supplied by the oil and gas industry [1]. It is also anticipated that by 2040, 57% of global energy demand will be fulfilled by the oil and gas industry. These numbers demonstrate the present and future impact of the oil and gas industry on our day-to-day lives. Although the use of oil as a source of energy dates back hundreds of years, the first oil well in modern history was drilled in Baku, Azerbaijan, in 1848 [2].

In 1859, Edwin Drake drilled the first commercial oil well in the United States (see Figure 1-1). Drake's well is a turning point for the modern oil and gas industry, which resulted in a great wave of investments in oil well drilling. For more than 150 years, the industry has constantly evolved and adopted new technologies. The oil and gas industry is probably among the few industries that have the distinction of fueling the industrial revolutions. Over the past few centuries, the world has witnessed three industrial revolutions.

© Yogendra Narayan Pandey, Ayush Rastogi, Sribharath Kainkaryam, Srimoyee Bhattacharya, and Luigi Saputelli 2020
Y. N. Pandey et al., *Machine Learning in the Oil and Gas Industry*,
https://doi.org/10.1007/978-1-4842-6094-4_1

- The first industrial revolution started around 1760 with the invention of the steam engine. During this era, steam-powered tools facilitated mechanical production in the factories, and steam engines revolutionized how people traveled.

- During the late nineteenth century to early twentieth century, scientific principles of production were brought in to the industries. This led to the mass production of products in the assembly lines. Gasoline-engine powered cars manufactured by the Ford Motor Company were among the most impressive examples of mass production at large scale, which happened during the second industrial revolution.

- With ENIAC, the first general-purpose computer was built in 1945, seeds of the third industrial revolution or digital revolution were sown. By the 1980s, progress made in semiconductor technology revolutionized industries all over the world, as computing moved from mainframe computers to personal computers. The arrival of the Internet connected the world, and progress in information technology ushered a new era of industrialization.

Figure 1-1. *Drake's well is considered the first well successfully drilled using a drilling rig [3]*

Currently, the industries across the globe are undergoing the fourth industrial revolution. This revolution is fueled by extraordinary growth in computing infrastructure, predictive techniques, data storage, and data processing capabilities. The fourth industrial revolution—sometimes denoted as the 4.0 version attached to the different industry names—is being fueled by the following enablers.

- Industrial Internet of Things (IIoT), which facilitates seamless connectivity between numerous devices in corporate and operational settings, and enables collaborative decision making.

- Big data technologies, which harness distributed data storage and processing to store and process enormous amounts of data efficiently.

- Cloud computing machines leveraging accelerated computing devices, such as graphics processing units (GPUs).

The oil and gas industry is also undergoing a digital transformation, which is referred to as "Oil and Gas 4.0." Oil and gas industry operations involve high-risk situations, both from human life, and environmental perspectives. As a result, the adoption of new technologies is possible only after a rigorous phase of validation to ensure that the health, safety, and environmental (HSE) standards are met appropriately. Despite the abundant caution exercised in new technology adoption, the oil and gas industry has remained open to new technologies for optimizing and streamlining the existing processes.

This book focuses on machine learning (a subset of artificial intelligence) applications in the oil and gas industry. In this chapter, without going into specific details of machine learning techniques, we discuss some example applications related to different life cycle stages of the upstream oil and gas industry. For the benefit of those of you with little background in the oil and gas industry, we start with a brief overview of the different industry sectors.

Major Oil and Gas Industry Sectors

Oil and gas deposits are often located thousands of feet below the earth's surface. The process of extracting oil and gas from these depths to the surface, and then converting it into a usable source of energy involves a large variety of operations. Figure 1-2 shows different life-cycle stages in the oil and gas industry operations. Broadly, the entire process of producing oil and gas is divided into the following three industry sectors.

- Upstream industry

- Midstream industry

- Downstream industry

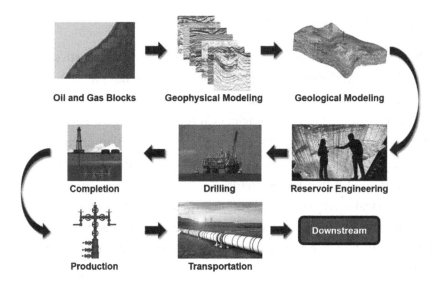

Figure 1-2. *Different life-cycle stages in a typical oil and gas industry operational environment*

The Upstream Industry

The upstream oil and gas industry is also known as the exploration and production, or E&P, industry. Operations in the upstream industry are focused on identifying locations below the earth's surface, which have the potential of producing oil and gas. Following the identification of a potential reserve, detailed planning of exploration, drilling oil wells, and producing oil and gas also comes under the upstream industry. Logically, upstream operations can be divided into the following activities.

- Exploration and appraisal

- Field development planning

5

- Drilling and completion

- Production operations

- Abandonment

Exploration and Appraisal

Finding the exact location of oil and gas deposits under the earth's surface is an activity as challenging as finding a needle in the haystack. Over the last century, scientific understanding of the earth's subsurface structure has improved significantly. Geological surveys provide geological maps or models of the structure under the earth's surface. In conjunction with geophysical surveys, these models help in generating a detailed understanding of subsurface structure.

Geophysical surveys can be of various types, depending on the techniques used for the survey, such as magnetic, electromagnetic, gravimetric, and seismic surveys, and so forth. In recent years, seismic surveys have proven to be the most reliable in locating potential oil and gas deposits. Seismic surveys can be conducted onshore (land) or offshore (marine).

Underlying principles of seismic imaging are similar to ultrasound imaging of the human body. The process of acquiring a seismic survey involves a source of acoustic energy. These sources can be large vibrator trucks on land and airguns in the case of a marine seismic. The acoustic waves reflected from different subsurface structures are received by geophones or hydrophones (very sensitive microphones) near the earth or water surface. 3D and 4D seismic surveys are among the most voluminous datasets in the oil and gas industry, sometimes reaching petabytes in size. The information from these seismic surveys helps identify subsurface structural features, and potential oil and gas deposits.

Geologists have a detailed understanding of the earth's depositional environment, which helps them understand how different rock layers (strata) are arranged under the earth's surface. Geologists use their deep knowledge of basin and sequence stratigraphy, along with the information from the interpretation of seismic data, to pinpoint locations where oil can be found.

Once potential locations of the petroleum reserves are identified with a certain level of confidence, the operating oil and gas company acquires a lease of the oil and gas block. Based on a combination of geophysical and geological understanding, exploratory wells (also called *wildcat wells*) are drilled to verify potential oil and gas reserve locations. If oil is found in the first wildcat well, reservoir engineers get involved in the field appraisal process.

During this process, more appraisal wells are drilled around the original wildcat well. Different well logs and core logs are gathered from the exploration wells, which are further used for developing a more detailed understanding of the discovered oil and gas reserve.

Geologists and petrophysicists use geological models along with the data from well logs to build detailed 3D reservoir models of physical properties in the region of interest, a process known as *petrophysical modeling*. Petrophysical properties, such as porosity (void fraction in the rocks, which can store or transport oil and gas), and oil and water saturation, are computed in the region of interest. These detailed petrophysical models help calculate a probabilistic estimate of the total available and recoverable volume of oil and gas in the discovered petroleum reservoir.

Field Development Planning

Once exploration and appraisal activities are over, reservoir engineers and geoscientists (geologists, and geophysicists) come together to develop a detailed plan for further development of the oil and gas producing field. All the information and insight gained from the exploration and appraisal phase—including the detailed reservoir models—are taken into

7

account. Decisions taken during this stage include identification of precise locations for drilling the wells, detailed design for optimal placement of the wells, design of well completion (including perforations and stimulation required).

Facilities engineers, reservoir engineers, and production engineers also develop a detailed facility design required to process the oil and gas (e.g., separators to isolate oil, water, and gas). A field development plan also includes the strategy for transporting produced oil and gas to storage or processing facilities. Once the field development plan is approved, the process of constructing facilities, drilling wells, and completing them as per the specifications provided in the field development plan begins.

Drilling and Completion

The process of drilling focuses on creating pathways for oil and gas to flow from the oil and gas reservoirs to the earth's surface. Drilling is a high risk and high capital expenditure (CAPEX) operation. Rental fees and other related charges for a deepwater offshore drilling rig may be close to US $1 million per day. As a ballpark figure, a single offshore well may cost from US $100 million to $200 million—or more in deep waters. The cost of drilling a well onshore (land) may range from US $5 million to $8 million [4].

During drilling operations, a drill bit is attached to the lowest end of the drill string, which is a long column of sequentially attached drill pipes made of steel or aluminum alloy. The top end of the drill string is attached to a drive (a mechanical device), which provides torque to the drill string.

As a result of this torque, the drill bit attached to the bottom of the drill string rotates and cuts through the rocks to create a wellbore. During the drilling operations, drilling mud is flown through the drill pipe to cool down the drill bit. Drilling mud also facilitates the removal of rock cuttings produced during the drilling process.

During the process of drilling, the drill string is retrieved periodically, and the wellbore is stabilized by cementing and lining with casing. The casing is high strength hollow steel pipe, which in conjunction with cementing, strengthens the wellbore, and keeps wellbore from collapsing. In addition to providing structural strength, the casing also ensures that chemical fluids used during the drilling process don't contaminate the nearby aquifers and groundwater sources.

Following the cementing, and running casing in the wellbore, production tubings are run into the wellbore. Tubings are hollow corrosion-resistant stainless-steel pipes always smaller in diameter than the casing, which are held inside the casing through packing devices. Tubing provides oil and gas passage from the bottom of the well (bottomhole) to the surface and protects the casing from corrosion and other possible wear and tear.

Once the casing and tubing are in place, perforations are created in the casing or casing and tubing to facilitate the flow of oil and gas from the reservoir rock into the wellbore. In addition to perforations, the completion activities may also include hydraulic fracturing and other production stimulation activities.

During hydraulic fracturing, water is injected inside the wellbore at very high pressure. The high pressure creates fractures in the rocks and creates additional pathways for oil and gas to flow from the reservoir to the wellbore. Proppants are small diameter sand particles, which are used during the fracturing process to ensure that fractures remain open. Proppant movement is supported by diverter, which are chemical compounds to facilitate the flow of proppant particles from the wellbore to inside the fractures.

Once all completion activities specified in the field development plan are completed, the well begins to produce oil and gas. It should be noted that activities mentioned in the last three pages may sometimes take five to ten years before the first drop of oil is produced from the well.

Production Operations

Once the well starts producing oil and gas, production engineers and reservoir engineers continue to monitor the production trends for the well. Throughout well's life span, the production of oil and gas from the well declines over time. Production engineers develop plans for maintaining oil flow from the well by using techniques, such as artificial lift. An artificial lift may utilize pumps (electrical submersible pumps, sucker rod pumps, etc.), or gas lift by injecting gas into the bottomhole to increase or maintain the flow of oil and gas from the well. During the production operations, other well intervention and workover activities, such as replacing a completion to maintain the flow from the well, can also be performed.

Abandonment

Through several years of continuous production, oil and gas production rates from the well keep declining. As mentioned earlier, engineers attempt to revive a well by well intervention and workover activities. However, after a certain point of time, it may be deemed uneconomical to revive a well given extremely poor production potential of the well. In such a scenario, well is plugged with cement and abandoned. This is the last stage in the life cycle of an oil and gas well. During the plugging and abandonment process, requirements set forth by different regulatory bodies for plugging the well are strictly followed to ensure compliance.

The Midstream Industry

The midstream industry is responsible for transporting crude oil produced from the oil wells to the downstream processing facilities. Transportation may take place via pipelines, oil vessels, trucks, or rails. In addition, the midstream industry is also responsible for the storage and marketing of petroleum commodities including, crude oil and natural gas. In this way, the midstream industry acts as a connecting link between the oil and gas production facilities in the remote areas, and the downstream processing facilities, which are often established close to populated areas.

The Downstream Industry

The primary functions of the downstream industry include fractionation or cracking of crude oil (a process of converting thick, high molecular weight crude oil to transparent looking, low molecular weight petroleum fuels) and related chemical treatments. This is followed by blending operations, which result in finished petroleum products for mass consumption as fuel. Besides, downstream refineries are also responsible for the purification of natural gas. The complete spectrum of the downstream industry includes oil refineries, petrochemical plants, distribution operations, retail centers, and other related facilities. Among the three sectors of the oil and gas industry, the downstream industry has the most visibility to the general population, as almost all of us use gasoline or diesel in our vehicles for our day to day commute. Jet fuel, asphalt, lubricants, plastics, and natural gas are among the other various petroleum products produced by the downstream industry.

So far, we have discussed the three sectors of the oil and gas industry. We deliberately kept the description of midstream and downstream industries short. In the next sections, we focus our discussion on digitalization of the upstream oil and gas industry and its connection with machine learning.

Digital Oilfields

The global oil and gas industry has an impressive economic footprint. Based on the data available in the public domain, worldwide revenue of the oil and gas companies was approximately US $5.5 trillion in 2017 [5].

Out of this, the top 10 companies shared almost half of this revenue (see Figure 1-3). In recent years, almost all these companies have started investing in digital transformation efforts. According to a report published by the World Economic Forum, digital transformation for the oil and gas industry may be worth from US $1.6 trillion to $2.5 trillion, between 2016 and 2025 [6]. These numbers give a glimpse of the sheer magnitude of oil and gas industry digitalization efforts.

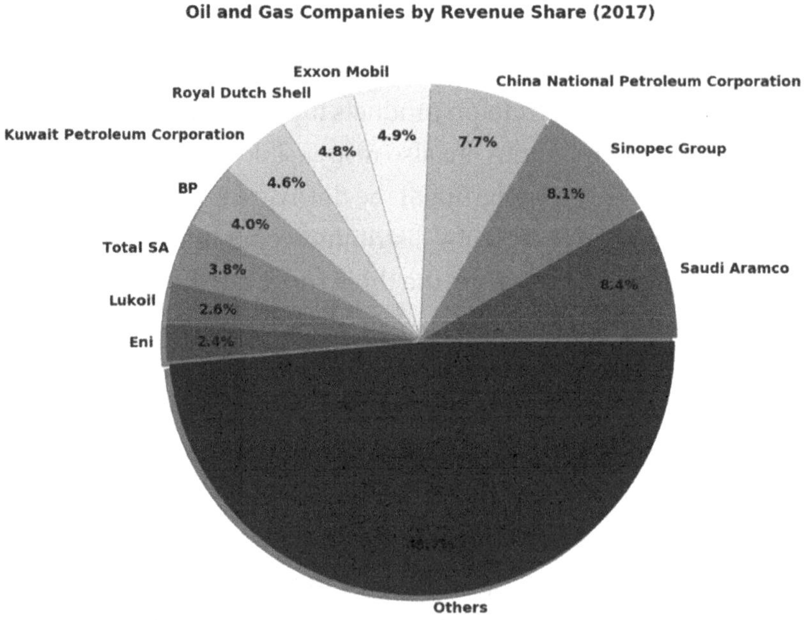

Oil and Gas Companies by Revenue Share (2017)

Figure 1-3. *World-leading oil and gas companies, and their share in the global revenue based on data for 2017 [5]*

Digitalization is not new to the oil and gas industry. Computer-assisted connected oil fields date to the late 1970s [7]. In the last two decades, the upstream oil and gas industry has faced an exponential increase in the use of real-time data, and field actuating devices, which have led to numerous digital oilfield implementations. These have demonstrated value to drive operational efficiency, optimize production, and maximize hydrocarbon

12

recovery factors. At the same time, digital oilfield implementations have facilitated better and faster decisions while reducing health, environmental, and safety risks. Formally, digital oilfield is a process of a measure-calculate-control cycle at a frequency which maintains the system's optimal operating conditions at all times within the time constant constraints of the system, while sustainably

- Maximizing production

- Minimizing capital expenditure (CAPEX) /operational expenditure (OPEX)

- Minimizing environmental impact

- Safeguarding the safety of the people involved and the integrity of the associated equipment [8] [9]

These objectives can be achieved by synchronizing four interrelated areas: people, automated workflows, processes, and technologies. It should be noted that significant efforts are required to properly describe a work process that supports digital oilfield implementation. In a simplified way, digital oilfields are a collection of automation and information technologies that transform the philosophy of the way petroleum assets are managed and drive more efficient operating work processes. It involves orchestration of disciplines, data, applications, and workflow integration tools supported by digital automation (see Figure 1-4), which involves the following.

- Field instrumentation

- Telemetry

- Automation

- Data management

- Integrated production models

- Workflow automation

- Visualization

- Collaboration environments

- Predictive analytics

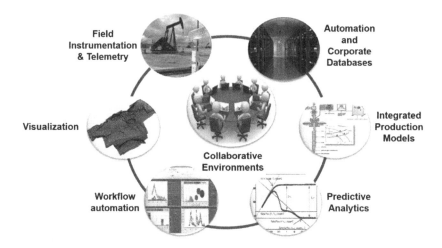

Figure 1-4. *Different components of a Digital Oilfield (DOF)*

A successful digital oilfield implementation requires synchronization among all of these components. The predictive analytics component of a digital oilfield is responsible for generating data-driven insight for the present and future prospects of the oilfields. This component often hosts machine learning algorithms and is a key to successful digital oilfield implementation. In the next section, we discuss how machine learning is providing predictive analytics capability to the digital oilfield implementations, and the oil and gas industry at large.

Upstream Industry and Machine Learning

The upstream industry has used machine learning for decades. However, it may have been referred to as artificial intelligence, or by other alternate names. Figure 1-5 shows a composite timeline of machine learning and oil and gas industry milestones. In the timeline, special emphasis has been given to the oil and gas industry milestones representing an advancement in data acquisition and digitalization. Based on the nature of advancements in the timeline, we can group the timeline into distinct periods. The time up to the first decade of the twentieth century is the foundation for both industries.

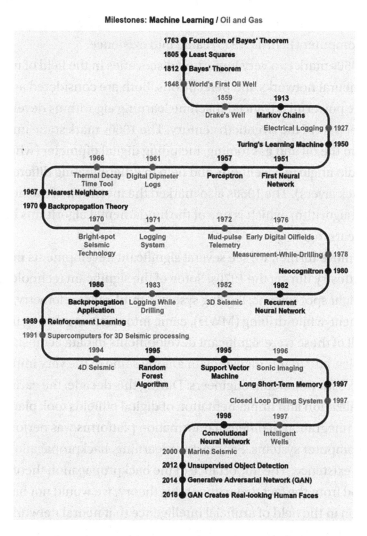

Figure 1-5. *Oil and gas industry and machine learning milestones. Compiled from references [2] and [10]*

During the early years, fundamental theoretical discoveries—such as Bayes' theorem and the least-squares method—took place in the machine learning arena. In the oil and gas industry, first commercial oil wells were drilled, and oil gained popularity as an important fuel, slowly replacing coal-fired steam engines. As we move toward the mid-twentieth century, we observe the use of electrical logging in the oil and gas industry. In contrast, Turing's learning machine appears shortly after the first general-purpose computer (ENIAC, 1945) came into existence.

The 1950s mark two very important discoveries in the field of machine learning: neural networks and perceptrons. Both are considered as the basis of the powerful advanced machine learning algorithms developed in the latter part of the twentieth century. The 1960s mark some important advances in the oil and gas tooling, including digital dipmeter (which could log dip angle in a wellbore, and help in differentiating different types of rock layers). The 1960s also marked the invention of the nearest-neighbors algorithm, which is one of the fundamental algorithms in machine learning.

As we move further, we see several significant developments in the oil and gas industry during the 1970s. Some of the significant technologies, such as bright spot seismic, logging systems, mud-pulse telemetry, and measurement-while-drilling (MWD), came into existence during this decade. All of these were significant developments in data acquisition technologies, and to this date, remain among some of the very important data sources for petroleum engineers. During this decade, the earliest conceptualization and implementation of digital oilfields took place, where the integration of different information platforms was performed through computer systems. For machine learning, backpropagation theory came into existence. The importance of the backpropagation theory can be understood from the fact that without this theory, we would not have seen a revolution in the field of artificial intelligence that neural networks have ushered in the recent years.

During the 1980s, 3D seismic forever changed the way geoscientists studied subsurface structures. Around the same time, logging-while-drilling (LWD) enabled access to real-time data during drilling operations, providing precise information about the physical properties of the subsurface formation. During the same decade, the invention of neocognitron laid down the foundation of advanced computer vision, and recurrent neural network (RNN) provided a new tool for sequence modeling. It can only be a coincidence that one of the most important advances in 3D seismic imaging of the subsurface, and discovery of the fundamental computer vision algorithm neocognitron, were a couple of years apart. In Chapter 4 of this book, you will see how these two events, which happened only two years apart in two different industries, can be connected to generate insight about the earth's subsurface structure by applying computer vision algorithms to the seismic images.

Two more essential discoveries in the field of machine learning during the 1980s were the application of backpropagation theory to facilitate the training of neural networks and the foundation of reinforcement learning. Reinforcement learning is one of the machine learning paradigms, which relies on reward functions to train software agents for accomplishing specific tasks. These are the algorithms that can teach a virtual player to win a game against a human, or teach a robot to walk like a human.[1]

In the 1990s, significant advancements happened in seismic technology. On one hand, supercomputers were employed to process 3D seismic data. On the other hand, 4D seismic data helped geoscientists interpret and visualize the evolution of the subsurface structures over time. This was also a decade of another beautiful coincidence. As geoscientists gained the capability of understanding the evolution of seismic in a time sequence, machine learning found a new algorithm called *long short-term memory*

---

[1]We do not cover these topics in this book, but please feel free to explore them at easily accessible resources, such as https://gym.openai.com, after you have a fundamental understanding of machine learning, which is covered in Chapter 3.

(LSTM) for sequence prediction. These two events are marked on the timeline a mere three years apart. LSTM addressed some limitations in the conventional recurrent neural networks and provided a better capability of sequence prediction. You will learn more about LSTMs in Chapter 3.

The 1990s also witnessed the discovery of random forest and support vector machine algorithms. Over time, the random forest algorithm has earned a place among the go-to algorithms of every machine learning practitioner.

The 1990s was a decade of accelerated research in the field of machine learning, and one of the fascinating algorithms to be published during this decade was the convolutional neural network (CNN). CNNs have been a driving force in the field of computer vision in the past decade. During the 1990s, the oil and gas industry also witnessed new technologies, such as sonic logging, closed-loop drilling systems, and intelligent wells.

Since the beginning of the twenty-first century, the oil and gas industry has seen significant technological advancement in the space of data acquisition. Along with the vast amounts of data acquired by the offshore seismic surveys, the boom in unconventional oil and gas industry has also been a source of large amounts of data. Furthermore, computing hardware advances, such as graphics processing units (GPUs) based computers, have accelerated computing capability significantly. Leveraging fast-computing hardware in conjunction with enormous amounts of training data, machine learning algorithms developed during the recent years have become capable of identifying objects in the images and videos without any human supervision. On the other side, a new class of algorithms, generative adversarial network (GAN) have been able to generate human faces, which are nearly impossible to distinguish from the real human pictures. Interestingly, GANs are also effectively used in seismic image reconstruction [11].

The milestones review of the two industries shows significant progress both industries have made collectively in data acquisition, digitalization, and capability of learning from the vast amounts of data. In the following part of this chapter, we are going to briefly review the use of machine learning in different parts of the upstream industry. First, we provide a brief overview of different types of data available in a specific upstream industry section, and then we discuss cases where this data is used with machine learning algorithms to solve upstream industry problems. For those of you without a background in machine learning, it is important to get a gentle introduction of these algorithms, before learning their applications in the oil and gas industry. Therefore, in the coming chapters of this book, we first go over the details of these machine learning algorithms. Then we show examples of how these algorithms are used for solving industry-specific problems. In this chapter, it is sufficient for you to understand that there are different kinds of machine learning algorithms, which have been applied to solve various types of industry-related problems.

Geosciences

Geosciences can be understood as a discipline comprising of fields of geology, geophysics, and petrophysics. In the following sections, we discuss a few problems related to geosciences, which have been solved by the application of machine learning algorithms.

Geophysical Modeling

Geophysicists work with the seismic data. Seismic data may be 2D, 3D, or 4D (depicting temporal evolution). Due to sheer volume, seismic datasets are the most suitable for advanced machine learning algorithms, such as CNN, which require a lot of data to train adequately. Some of the problems in geophysics that can be solved by the application of machine learning are discussed here.

Automated Fault Interpretation

Fault interpretation is an essential step in seismic interpretation. Faults are formed by different mechanisms of subsurface movements, which result in a structural discontinuity. Certain operational decisions need to be taken after considering the presence of faults in the region of interest. In conventional workflows, fault interpretation is a time-intensive process. Convolutional Neural Networks may be instrumental in formulating and solving fault interpretation as a computer vision problem. However, it may not always be possible to generate large amounts of labeled data representing seismic images, and corresponding fault locations to train the CNN models. However, recent research has established methodologies for training advanced CNN variants with minimal training datasets, while providing highly accurate pixel-by-pixel prediction of the fault locations in the provided seismic images [12] [13].

Automated Salt Identification

Another crucial task in seismic interpretation is to identify the salt bodies in the subsurface environment. In the conventional workflows, visual picking of the salt/sediment boundaries is the standard way of performing salt identification. This is a time-consuming manual process, which may be affected by individual bias. Advanced CNN-based classification, along with the residual learning, has demonstrated high precision in delineating salt bodies. The results from these CNN-based algorithms show good agreement with manually interpreted salt bodies. Such results indicate the potential of applying CNN for automated salt identification [14].

Seismic Interpolation

Acquisition of high resolution, regularly sampled seismic data is often hindered due to physical or financial constraints.

This leads to undersampled seismic data sets. Also, the presence of data quality issues, such as bad or dead traces, introduces additional challenges in the exploration and production activities that depend on high resolution and high-quality seismic images. Addressing these challenges can enable geoscientists, reservoir engineers, and other industry professionals to improve the predictive capability of their models.

In recent years, generative adversarial networks have been employed for the interpolation of seismic images. Seismic interpolation helps in reconstructing the bad, and dead traces. It may also be used as a resolution enhancement tool for seismic data. The results of seismic interpolation using a generative adversarial network make it an interesting alternative to classical methods [11].

Seismic Inversion

The process of seismic inversion transforms seismic data into quantitative rock property estimates describing the reservoir. In recent research, GAN has been combined as an *a priori* model for creating subsurface geological structures, and their petrophysical properties. The generated structural and properties information was used for assisted seismic inversion [15]. This is another interesting application of GANs in the field of seismic processing.

Geological Modeling

Geological modeling is the process of generating detailed 2D or 3D models of physical properties in the region of interest, at and below the surface of the earth. The modeling process uses geological observations at the earth's surface, data available from different types of well logs from the wells drilled in the region, and seismic data. Some machine learning applications in geological modeling are being discussed here.

Petrophysical Modeling

Petrophysical modeling is critical for gaining understanding about the structure and properties of zones of oil and gas deposits in a reservoir. Neural networks have been successfully applied to create a detailed representation of petrophysical property distributions in the oil and gas reservoirs, which is sometimes called a *neural kriging* [16] [17]. During recent years, deep neural networks have been applied for generating detailed 2D and 3D models or reservoirs [18] [19]. These studies demonstrate the effectiveness of machine learning methodologies in constructing detailed models of petrophysical properties.

Facies Classification

Facies can be understood as a class of rock with distinct characteristics. Understanding of facies distribution in an oil and gas reservoir can help in identifying the regions, which are more likely to have oil and gas producing rocks. In the recent years, machine learning approaches applying deep neural networks have shown reasonable accuracy in classifying facies based on the petrophysical properties from the well logs [20], and at random points of interest away from the wells [19].

Reservoir Engineering

Reservoir engineers have access to almost all the data used by geoscientists. They also have hydrocarbon production and operational history data for the field they are responsible for. They use a diverse variety of data to develop a plan for economically optimal exploitation of oil and gas from the reservoir, and the whole oilfield at large. Some of the reservoir engineering activities, where machine learning has shown potential benefits, are discussed next.

Field Development Planning

During the field development planning, reservoir engineers use data from the existing wells. Based on the analysis of data from existing wells, they decide the placement of new wells. Specifically, this is a standard practice in the development of unconventional oil and gas resources. There have been several approaches to applying machine learning to plan new wells based on historical data.

One such approach used dimensionality reduction by applying principal component analysis (PCA), followed by regression methods to predict the production potential of new proposed wells [21]. The analysis based on machine learning methods has also demonstrated the capability of uncovering hidden patterns, which are not easily noticed in a high dimensional space. By identifying these patterns, it's demonstrated that the wells expected to behave similarly can be identified. Further analysis can reveal common properties of wells exhibiting similar behavior, the underlying reasons for poor performance in the wells not behaving satisfactorily, and measures to avoid similar poor performance in the new wells [22].

Assisted History Matching

The process of history matching follows an iterative scheme for tuning reservoir parameters. The process converges when a reasonable match between the history-matched and observed reservoir performance is achieved. In the reservoirs with high heterogeneity, and complexity, history matching may be a time-consuming process with a high computational cost. In recent years, artificial neural network-based methodologies have been applied to reservoir history matching [23]. Such methodologies employ fully data-driven proxy models, which reduce the computational cost significantly. The proxy reservoir models run at a fraction of computational cost when compared to a full-scale numerical reservoir simulator. This reduction in computational cost makes the process of history matching highly efficient [24].

Production Forecasting and Reserve Estimation

Forecasting oil production from the oil-producing wells is among the main responsibilities of reservoir engineers. Based on the generated production forecast, decisions about future field development activities are taken. Production forecasting helps in the estimation of available reserves and economic evaluation. Traditionally numerical simulations and decline curve analysis (DCA) techniques are extensively used for production forecasting. These techniques require knowledge of the reservoir behavior and lack flexibility for modeling complex physics.

In recent years, machine learning methodologies using feed-forward neural networks [25] [26], and recurrent neural networks have demonstrated their effectiveness and accuracy in a production forecast of single and multiple wells. These machine learning approaches make the process of forecasting efficient and accurate for the assets with or without significant operational history information [27].

Drilling and Completion

Offshore drilling is a high-risk and high-CAPEX operation. Deepwater offshore drilling may cost as much as US $1 million per day, and a single well may cost hundreds of millions of dollars in a deepwater drilling scenario. Provided these facts, automation employing machine learning models can improve operational safety in the drilling operations significantly, and bring the drilling costs down simultaneously. Some of the applications of machine learning in the drilling and completion activities, which demonstrate the significant promise of improved operational safety, and cost benefits, are discussed next.

Automated Event Recognition and Classification

Upstream assets create valuable insights from manually generated operator reports and sensor data, such as weight on bit, standpipe pressure, rotation frequency, and so forth. Such value can be achieved if and only if data is of adequate quality and information is properly interpreted, stored, and used. Reports and data provide the basis for deriving well events. Examples of well events include productive time (e.g., drilling hole), non-productive time or NPT (e.g., stuck pipe, stick-slip vibration, hole cleaning, pipe failures, loss of circulation, bottomhole assembly whirl, excessive torque and drag, low rate of penetration, bit wear, formation damage, and borehole instability, etc.), and invisible lost time or ILT (e.g., slow pipe connection, suboptimal drilling, and tripping).

For many years, drilling and completion operations have been supported with extensive data collection processes. In the last two decades, with the inclusion of downhole sensors and telecommunication capabilities, there is a broad spectrum of real-time data collection processes that occur during drilling and completion operations. The current challenge is to identify events from real-time data and derive insights to prevent NPT and ILT. Traditionally, daily drilling reports (DDR) have been the source of truth for describing detailed log of key operations and events occurring at the rig site; however, they are subject to human bias, and they use inconsistent text formats and nomenclature. The objective of automated event recognition and classification is to pinpoint and rectify problems that occur in the drilling process [28] and within a time frame that enables optimum and safe drilling operation.

Various machine learning techniques have been proposed to automatically identify and classify drilling events by detecting the changing trend of drilling parameters when the events happen. Drilling events are extracted from massive DDR and real-time drilling databases using defined expert rules, thresholds, and criteria [29], as well as regression and classification methods [30]. There have also been attempts to maximize the value from DDR and real-time data [31].

Non-Productive Time (NPT) Minimization

Non-productive time (NPT) in the drilling operations may arise out of multiple reasons. Due to the high rental fees of drilling equipment, there's an economic incentive for minimizing such non-productive time. Kicks and stuck pipe are among two of the most significant problems leading to non-productive time. A machine learning approach to predict the incidence of these problems is discussed next.

Early Kick Detection

Kick is an event where gas suddenly starts to seep in from the formation into the wellbore. As gas enters the wellbore, it starts ascending to the surface. The upward movement of gas shows up as the increased volume of mud at the surface, accompanied by increased mud flow rate out of the well.

Kick is a dangerous event, and remedies to the aftereffects of such events prove to be expensive. The earlier a kick is detected, the sooner the crew on drilling rig can take necessary corrective actions. However, traditional alarm systems either suffer from many false alarms or are ineffective at identifying a kick accurately. The application of machine learning algorithms provides accurate estimates of mud volumes and flow rates during the drilling process. It also significantly reduces the frequency of false alarms. This helps the drilling crew in taking effective corrective actions promptly [32].

Stuck Pipe Prediction

In certain scenarios during the drilling operations, the drill pipe may get stuck inside the wellbore. Once the drill pipe gets stuck, drilling operations are put on hold until a remedy is performed. Stuck pipe problem is one of the biggest contributors to non-productive time during drilling operations. To predict the nonlinear behavior surrounding the stuck pipe events, artificial neural networks optimized by particle swarm optimization algorithm have shown high accuracy in predicting the possibility of a stuck pipe event ahead of time. Predictions obtained from this machine learning approach, together with sound engineering judgments, has the potential of preventing stuck pipe events [33].

Autonomous Drilling Rigs

Large offshore drilling rigs are often located in remote locations and can generate up to 1 or 2 terabytes of data every day. Transmitting such large amounts of data generated by an offshore rig in one day using a 2 Mbps satellite link takes 12 days or more [34]. Therefore, any critical data-driven decision making needs to take place on the offshore drilling rig itself. Autonomous drilling rigs combine machine learning, IIoT, and robotics. Also, the decision making on the rig needs edge computing devices (devices with electronic circuits embedding machine learning and other algorithms) installed at the rig. The sophisticated machinery and a large number of sensors on the drilling rig provide all the relevant information for decision making. The data acquired by the IIoT devices can be consumed by machine learning algorithms running on the edge computing devices. Finally, recommendations from the algorithms transmitted back to drilling operators or robots result in the final action. The objectives of autonomous drilling rigs include the following.

- Drilling at optimal rates (rate of penetration or ROP optimization) [35].

- Minimizing non-productive time (NPT), for example, predicting the possibility of stuck pipe ahead of time, and increasing probability of success of different remedial actions (25% of total NPT during drilling operations is attributed to stuck pipes) [33].

- Operational safety, for example, early kick detection to ensure the safety of human lives and preserving the ecosystem [32].

Production Engineering

As oil and gas fields mature, reservoir and well conditions change continuously, and most likely toward deterioration. As a result, production declines rapidly due to an increase in water cut (fraction or percentage of water present in oil), lift equipment inefficiencies, lack of sufficient reservoir pressure support, or surface backpressure bottlenecks, among many others. Reservoir and field decline may be a result of individual well production decline, as well as the rapid increase of inactive strings (non-producing wells).

Drilling new wells or the intervention of existing ones are needed to manage production decline. To improve production through an existing well, reservoir and production engineers plan and execute workover jobs consisting of squeeze cement, well integrity, fishing jobs, and zone isolation to optimize and maintain oil production.

The role of production engineering is to ensure reservoir production delivers within the existing boundaries of the approved field development plan, including the number of wells, facilities capacity, and limited expenditures. The responsibilities of a production engineer may include the following.

- Plan, gather, and interpret production surveillance data.

- Build, maintain, and exploit the use of models for well performance management.

- Establish well-operating envelopes and set points.

- Identify, rank, implement, and assess production enhancement opportunities, including well workovers.

Workover Opportunity Candidate Recognition

Under certain scenarios during its lifetime, a well may not be able to operate and produce satisfactorily. Identifying such wells, and performing interventional and remedial activities (also called *workover*) to bring the well performance to satisfactory levels has shown significant economic impact. Expert systems have been used in the oil and gas industry for decades, with processes ranging from data integration and cleansing over problem detection to ranking candidates according to production gains, costs, risk, and net present value. Novel systems leveraging data analytics, machine learning and reasoning tools like Bayesian Belief Networks [36] [37] have been proposed and used to detect production problems with likelihoods of occurrence for various problems in wells, reservoirs or facilities. These systems assist in differentiating root causes and prioritizing different countermeasures at hand. These screening logics work as a repeatable and automated process, which can be scheduled or can be executed on demand.

Production Optimization

The objective of production optimization is to manage and ensure asset goals profitably using all available information up to that point to predict outcomes with confidence and to make decisions that produce optimal outcomes, and implement such decisions until the next decision-making point in time. The increasing availability of real-time downhole measurements and remotely activated valves in the oilfields has made field-wide optimization of operations in real-time a distinct possibility [38]. With more real-time data and measurements, it's possible to build machine learning models, which could be updated with the availability of new data. While the term real-time optimization (RTO) is certainly not new, and RTO is practiced in elements of production operations, the extent to which RTO is now feasible in the day to day production operations has increased dramatically.

Infill Drilling

Infill drilling refers to additional drainage locations that are selected in a later stage of the field development cycle to increase the well-reservoir contact area. The objective of infill drilling is to accelerate or improve recovery factors from wells, which might have bypassed oil and gas reservoirs in their drainage area. Selecting infill well candidates requires a careful analysis to verify, quantify, and locate such bypassed reservoir locations. Numerous data-driven methods have been developed to forecast production for potential sidetracks providing a set of criteria to select the most suitable well to sidetrack, leveraging all associated uncertainties and linking to a stochastic economic analysis [39].

Optimal Completion Strategy

Completion is a term to indicate the hardware deployed inside the wellbore to ensure production lifting from the reservoir to the wellhead. Optimal completion strategy requires the proper understanding of reservoir potential (i.e., multiphase fluids, solids, productivity, and pressure regimes as a function of time) as well as the production requirements (i.e., rate targets and wellhead pressure).

Many authors use multivariate analysis, which includes multiple linear regression analysis and statistical models using multidisciplinary data, such as petrophysical (e.g., thickness, permeability, saturation), completion (e.g., tubing size, stimulation fluid, and fracture intensity), and production (e.g., pressure and multiphase flow tests data). Traditionally, production engineering models, such as nodal analysis well models are good enough to provide accurate predictive performance relationships for successful completions. In the cases of unknown well failure or uncertain reservoir phenomena, data-driven models provide much better prediction than pure engineering models.

Over the last five years, there has been an explosion in the applications of machine learning algorithms for predicting completion performance in unconventionals [40] [41]. Data-driven methods assist operators in the selection of optimum completion parameters, which have a positive effect on production and are advantageous to lower unit costs, such as stage intensity, plug-and-perforate cemented-well designs, injection rate, and proppant mass per lateral foot and fluid volume per lateral foot [42].

Predictive Maintenance

The upstream oil and gas industry employs a vast array of equipment. All equipment have an operational lifetime, before the ultimate failure. More than 90% of all producing oil wells need an artificial lift method to increase oil production. The *electrical submersible pump* (ESP) is widely used for artificial lift, making it a critical component in ensuring continued oil production from the wells.

Despite being among critical equipment in the upstream operations, ESPs exhibit significantly high failure rates [43] [44]. These failures are often random and result in lost oil production from the wells. It is estimated that ESP failures lead to hundreds of millions of barrels of lost or deferred oil production each year. Previous studies have used a *principal component analysis* (PCA) approach to detect ESP failures and predict the remaining useful life of the equipment before failure by using complete historical data. An imminent failure often shows up as PCA data scattered away from the origin. Combining these predictions from the machine learning-based models with engineering principles to detect problems with ESPs, long before they occur, and prescribing preventive actions can have a significant economic impact [45] [46].

Industry Trends

So far, we have discussed several applications of machine learning in the upstream oil and gas industry. Now we analyze some trends in the oil and gas industry and machine learning industry at large by analyzing research publication trends (see Figure 1-6).

For analyzing the industry trends, we used different search queries on Google Scholar to understand how publication trends in the machine learning community within the oil and gas industry and the general machine learning community have been evolving. The following interesting trends have emerged.

- There was a crossover in the dominance of the term *artificial intelligence* over the term *machine learning* in the general research community in 2008. A similar crossover was observed in the oil and gas community in 2018. This indicates that the oil and gas industry has adopted a machine learning lexicon with a delay of ten years.

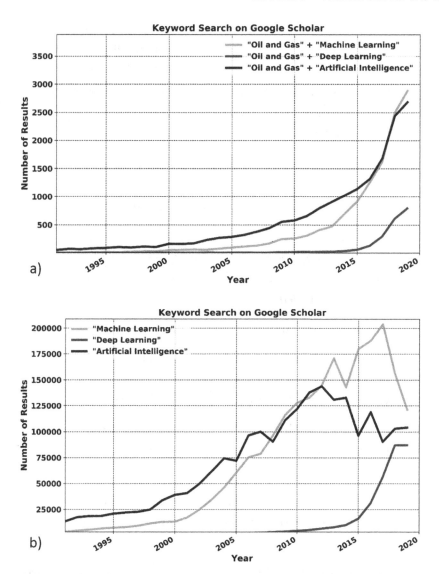

Figure 1-6. *Trends from 1991 to 2019, based on the number of Google Scholar keyword search results for a) oil and gas industry, and b) all industries related machine learning, deep learning, and artificial intelligence publications. Citation count for 2019 is estimated based on data from March 2019*

- After 2012, there was a steep decline in the popularity of the term *artificial intelligence* in the general research community. This decline fueled growth in the use of the terms *machine learning* and *deep learning*. The year 2019 may have marked the beginning of a similar trend for the oil and gas industry.

- The pattern observed for the general research community in 2010 for *deep learning* began to appear for the oil and gas industry around 2015, almost a five-year delay.

- For the general research community, *machine learning* showed a steep decline in 2017. Around the same time, the rise of *deep learning* continued.

Based on these observations, the following conclusions may be drawn.

- Adoption of machine learning and deep learning in the oil and gas industry has lagged by around five to ten years when compared with the general industry.

- Currently, the adoption of machine learning and deep learning in the industry is on the rise.

- In the near future, growth in machine learning and deep learning should continue. At some point, deep learning will take over machine learning in terms of popularity.

If you are wondering whether it is still worth investing time to learn machine learning rather than deep learning, the answer is yes. Deep learning is a subfield of machine learning. In Chapter 3, we will discuss the fundamentals of both machine learning and deep learning. There is a significant potential for growth in the adoption rates of these techniques in the oil and gas industry.

Summary

In this chapter, we provided an overview of the different segments of the oil and gas industry. We also compared the timeline of data acquisition and integration technologies in the upstream oil and gas industry, with the historical evolution of machine learning algorithms. Further, we provided an overview of applications of machine learning to some upstream industry problems. We hope that you now have a general understanding of the upstream oil and gas industry, where machine learning can provide valuable solutions. In the coming chapters, we will select some of these problems and see how to build machine learning solutions to these problems using real data sets. Before that, the next couple of chapters will be dedicated for overview of Python programming, machine learning, and deep learning concepts.

References

[1] ExxonMobil, *Outlook for Energy*, 2018.

[2] Society of Petroleum Engineers, "History of Petroleum Technology," [Online]. Available: https://www.spe.org/industry/history/timeline/.

[3] Wikipedia, "Drake Well," [Online]. Available: https://en.wikipedia.org/wiki/Drake_Well.

[4] American Geosciences Institute, "Subsurface Data in the Oil and Gas Industry," [Online]. Available: https://www.americangeosciences.org/critical-issues/factsheet/pe/subsurface-data-oil-gas-industry.

[5] Wikipedia, "List of largest oil and gas companies by revenue," [Online]. Available: https://en.wikipedia.org/wiki/List_of_largest_oil_and_gas_companies_by_revenue.

[6] World Economic Forum, "White Paper: Digital Transformation Initiative: Oil and Gas Industry," 2017.

[7] R. A. Startzman, W. M. Brummett, J. C. Ranney, A. S. Emanuel, and R. M. Toronyi, "Computer Combines Offshore Facilities and Reservoir Forecasts," *Petroleum Engineer,* pp. 65–76, May 1977.

[8] S. Mochizuki, L. A. Saputelli, C. S. Kabir, R. Cramer, M. Lochmann, R. Reese, L. Harms, C. Sisk, J. R. Hite, and A. Escorcia, "Real-Time Optimization: Classification and Assessment," *SPE Production & Operations,* vol. 21, no. 4, pp. SPE-90213-PA, 2006.

[9] L. Berendschot, K.-C. Goh, M. Stoever, R. Cramer, and S. Mehrotra, "Upstream/Downstream Real-Time Surveillance and Optimization Systems: Two Sides of the Same Coin, or Never the Twain Shall Meet," SPE Annual Technical Conference and Exhibition, San Antonio, TX, 2013.

[10] Wikipedia, "Timeline of machine learning," [Online]. Available: `https://en.wikipedia.org/wiki/Timeline_of_machine_learning`.

[11] D. A. B. Oliveira, R. S. Ferreira, R. Silva, and E. V. Brazil, "Interpolating Seismic Data With Conditional Generative Adversarial Networks," *IEEE Geoscience and Remote Sensing Letters,* vol. 15, no. 12, pp. 1952–1956, 2018.

[12] S. Li, C. Yang, H. Sun, and H. Zhang, "Seismic Fault Detection Using an Encoder-Decoder Convolutional Neural Network with a Small Training Set," *Journal of Geophysics and Engineering,* vol. 16, no. 1, p. 175–189, 2019.

[13] X. Wu, Y. Shi, S. Fomel, and L. Liang, "Convolutional Neural Networks For Fault Interpretation in Seismic Images," SEG International Exposition and 88th Annual Meeting, Anaheim, CA, 2018.

[14] Y. Zeng, K. Jiang, and J. Chen, "Automatic Seismic Salt Interpretation with Deep Convolutional Neural Networks," The Third International Conference on Information System and Data Mining, 2018.

[15] L. Mosser, O. Dubrule, and M. J. Blunt, "Stochastic Seismic Waveform Inversion Using Generative Adversarial Networks As A Geological Prior," *Mathematical Geosciences*, November 2018.

[16] D. M. Rizzo and D. E. Dougherty, "Characterization of Aquifer Properties Using Artificial Neural Networks: Neural Kriging," *Water Resources Research,* vol. 30, no. 2, pp. 483–497, 1994.

[17] S. Mohaghegh, R. Arefi, S. Ameri, K. Aminianda, and R. Nutter, "Petroleum reservoir characterization with the aid of artificial neural networks," *Journal of Petroleum Science and Engineering,* vol. 16, no. 4, pp. 263–274, 1996.

[18] M. Korjani, A. Popa, E. Grijalva, S. Cassidy, and I. Ershaghi, "A New Approach to Reservoir Characterization Using Deep Learning Neural Networks," SPE Western Regional Meeting, Anchorage, AK, 2016.

[19] Y. N. Pandey, K. P. Rangarajan, J. M. Yarus, N. Chaudhary, N. Srinivasan, and J. Etienne, "Deep Learning-Based Reservoir Modeling," Patent PCT/US2017/043228, July 21, 2017.

[20] B. Hall, "Facies Classification Using Machine Learning," *The Leading Edge,* vol. 35, no. 10, pp. 906–909, 2016.

[21] S. Bhattacharya and M. Nikolaou, "Using Data From Existing Wells To Plan New Wells in Unconventional Gas Field Development," Canadian Unconventional Resources Conference, Calgary, Alberta, Canada, 2011.

[22] S. Bhattacharya and M. Nikolaou, "Analysis of Production History for Unconventional Gas Reservoirs With Statistical Methods," *SPE Journal,* vol. 18, no. 5, 2013.

[23] L. C. Reis, "Risk Analysis with History Matching Using Experimental Design or Artificial Neural Networks," SPE Europec/EAGE Annual Conference and Exhibition, Vienna, Austria, 2006.

[24] A. Shahkarami, S. D. Mohaghegh, V. Gholami, and S. A. Haghighat, "Artificial Intelligence (AI) Assisted History Matching," SPE Western North American and Rocky Mountain Joint Meeting, Denver, CO, 2014.

[25] S. D. Mohaghegh, O. S. Grujic, S. Zargari, and A. K. Dahaghi, "Modeling, History Matching, Forecasting and Analysis of Shale Reservoirs performance Using Artificial Intelligence," SPE Digital Energy Conference and Exhibition, Woodlands, TX, 2011.

[26] J. Rebeschini, M. Querales, G. A. Carvajal, M. Villamizar, F. M. Adnan, J. Rodriguez, S. Knabe, F. Rivas, L. Saputelli, A. Al-Jasmi, H. Nasr, and H. K. Goel, "Building Neural-Network-Based Models Using Nodal and Time-Series Analysis for Short-Term Production Forecasting," SPE Middle East Intelligent Energy Conference and Exhibition, Manama, Bahrain, 2013.

[27] J. Sun, X. Ma and M. Kazi, "Comparison of Decline Curve Analysis DCA with Recursive Neural Networks RNN for Production Forecast of Multiple Wells," SPE Western Regional Meeting, Garden Grove, CA, 2018.

[28] F. E. Finch, G. M. Stanley, and S. P. Fraleigh, "Using the G2 Diagnostic Assistant for Real-Time Fault Diagnosis," European Conference on Industrial Applications of Knowledge-Based Diagnosis, Italy, 1991.

[29] C. I. Noshi and J. J. Schubert, "The Role of Machine Learning in Drilling Operations; A Review," SPE/AAPG Eastern Regional Meeting, Pittsburgh, PA, 2018.

[30] J. Zhao, Y. Shen, W. Chen, Z. Zhang and S. Johnston, "Machine Learning-Based Trigger Detection of Drilling Events Based on Drilling Data," SPE Eastern Regional Meeting, Lexington, KY, 2017.

[31] F. Zausa, S. Masi, J. Michelez, and N. Rossi, "Advanced Drilling Time Analysis Through the Combination of Operations Reporting & Sensors Data," Offshore Mediterranean Conference and Exhibition, Ravenna, Italy, 2013.

[32] S. Unrau, P. Torrione, M. Hibbard, R. Smith, L. Olesen, and J. Watson, "Machine Learning Algorithms Applied to Detection of Well Control Events," SPE Kingdom of Saudi Arabia Annual Technical Symposium and Exhibition, Dammam, Saudi Arabia, 2017.

[33] H. Toreifi, A. Manshad, H. Rostami, and A. Mohammadi, "Prediction and Elimination of Drill String Sticking Using Artificial Intelligence Technique," *Heavy Oil*, New York, NY: Nova Science Publishers, 2017, pp. 221–241.

[34] Cisco Canada Blog, "Beyond the Barrel: How Data and Analytics will become the new currency in Oil and Gas," [Online]. Available: https://gblogs.cisco.com/ca/2018/06/07/beyond-the-barrel-how-data-and-analytics-will-become-the-new-currency-in-oil-and-gas/.

[35] B. Mantha and R. Samuel, "ROP Optimization Using Artificial Intelligence Techniques with Statistical Regression Coupling," SPE Annual Technical Conference and Exhibition, Dubai, UAE, 2016.

[36] G. Zangl, L. Neuhofer, D. Zabel, P. Tippel, C. Pantazescu, V. Krcmarik, L. Krenn and B. Hachmöller, "Smart and Automated Workover Candidate Selection," SPE Intelligent Energy International Conference and Exhibition, Aberdeen, Scotland, UK, 2016.

[37] M. Stundner, G. Zangl, L. Neuhofer, D. Zabel, P. Tippel, C. Pantazescu, V. Krcmarik, A. I. Staicu, L. Krenn, and B. Hachmöller, "Deployment of a Generic Expert System to Rank Operations Business Opportunities Automatically Under Ever-Changing Economic Conditions," SPE Annual Technical Conference and Exhibition, Dubai, UAE, 2016.

[38] L. Saputelli, M. Nikoalou, and M. J. Economides, "Self-Learning Reservoir Management," SPE Annual Technical Conference and Exhibition, Denver, CO, 2003.

[39] P. W. von Pattay, J. A. Hamer, and R. Strasser, "Unlocking the Potential of Mature Fields: An Innovative Filtering and Analysis Approach to Identify Sidetracking Candidates in Mature Water Flooded Fields," Asia Pacific Oil and Gas Conference and Exhibition, Jakarta, Indonesia, 2007.

[40] B. L. Williams, B. M. Weaver, and L. Weijers, "Completing the Second Target: Finding Optimal Completion Practices for the Three Forks Formation in the Williston Basin Using Multivariate Statistical Analysis," SPE Hydraulic Fracturing Technology Conference, Woodlands, TX, 2015.

[41] F. Miller, J. Payne, H. Melcher, J. Reagan, and L. Weijers, "The Impact of Petrophysical and Completion Parameters on Production in the Denver-Julesburg Basin," SPE Low Perm Symposium, Denver, CO, 2016.

[42] M. Mayerhofer, O. Oduba, K. Agarwal, H. Melcher, E. Lolon, J. Bartell, and L. Weijers, "A Cost/Benefit Review of Completion Choices in the Williston Basin Using a Hybrid Physics-Based-Modeling/Multivariate-Analysis Approach," *SPE Production & Operations,* vol. 34, no. 01, pp. 1–17, 2019.

[43] R. Lastra, "Achieving A 10-Year ESP Run Life," SPE Electric Submersible Pump Symposium, Woodlands, TX, 2017.

[44] R. Pragale and D. D. Shipp, "Investigation of Premature ESP Failures and Oil Field Harmonic Analysis," *IEEE Transactions on Industry Applications,* vol. 53, no. 3, pp. 3175–3181, 2017.

[45] S. Gupta, L. Saputelli, and M. Nikolaou, "Applying Big Data Analytics to Detect, Diagnose, and Prevent Impending Failures in Electric Submersible Pumps," SPE Annual Technical Conference and Exhibition, Dubai, UAE, 2016.

[46] M. Abdelaziz, L. Rafael, and J. J. Xiao, "ESP Data Analytics: Predicting Failures for Improved Production Performance," Abu Dhabi International Petroleum Exhibition & Conference, Abu Dhabi, UAE, 2017.

CHAPTER 2

Python Programming Primer

In the previous chapter, we discussed some challenges in the oil and gas industry, where machine learning can provide data-driven insight and solutions. However, before you can implement those solutions, you need to learn how to code the applicable machine learning algorithms. This makes understanding a computer programming language necessary before diving into machine learning.

There are several programming languages that are used for developing machine learning solutions and applications across industries. Python, R, Java, C/C++, Julia, Scala, Go, and Lua are among some of the leading languages used for developing machine learning algorithms and applications. Out of these languages, Python had the largest community of machine learning practitioners as of 2019. Will this change over time? Perhaps another programming language will overtake Python in terms of popularity in the distant future. We refrain from getting into a debate about which programming language is better or worse, as that is not the intention of this book. It should be acknowledged that each programming language has its own merits and demerits. However, it makes practical sense to learn Python when working with machine learning algorithms, given the rich machine learning ecosystem and vibrant community that Python offers.

© Yogendra Narayan Pandey, Ayush Rastogi, Sribharath Kainkaryam, Srimoyee Bhattacharya, and Luigi Saputelli 2020
Y. N. Pandey et al., *Machine Learning in the Oil and Gas Industry*, https://doi.org/10.1007/978-1-4842-6094-4_2

Python has the following features.

- **Platform independent**. Python code written on one platform can run on another platform. It means, if you wrote code on Windows operating system, you do not need to rewrite it if you wanted to run it on the Linux operating system. Python has been ported to almost all commonly used platforms. However, there may be some minor tweaks needed for functionalities related to the operating system and file system.

- **Interpreted**. You do not need to compile the Python code explicitly before running it. Python code is executed by an interpreter, which takes care of converting Python code to byte code.

- **High-level**. While writing a Python code, you do not need to worry about intricate system-level details, such as how to allocate or free memory during program execution. Python provides an abstraction for a large variety of general-purpose functions so that you can focus more on programming logic than the system or hardware-level details.

- **Open source**. Python is freely distributed, even for commercial purposes. Anyone can see and make changes to Python source code. A large community of developers is constantly contributing new functionalities and features to the rich Python ecosystem for general-purpose computing, and machine learning.

- **Object-oriented**. Python supports an object-oriented programming paradigm, where you can define data structure and behavior of real-world entities using a set of variables and functions. We provide more details about this later in this chapter.

- **Dynamically typed**. When you define a variable in Python, you don't need to worry about defining its data type. In simple words, you don't need to describe whether a variable is an integer or a real number with decimals (float). Python dynamically assigns data types to variables.

- **Conversational syntax**. Python syntax is intuitive and conversational. Just to give an example, if you want to see if a number exists in a list of numbers, you may write if 5 in [1, 2, 3, 4, 5], and the result is True.

- **Easy to learn**. The features of Python make it an easy-to-learn programming language. While mastering Python may take significant effort, getting started with Python has a shorter learning curve as compared to most other programming languages mentioned earlier in this chapter.

Before we can start writing code in Python, we need to install Python and set up the programming environment. The next section walks you through the process of installing Python on your computer.

Installation and Environment Setup

There are many options to choose from when it comes to selecting an installer for Python. For this book, we recommend getting the most recent Python 3.x installer for the operating system installed on your computer, from the Anaconda distribution download website [2]. Anaconda distribution has most of the required Python libraries, and it comes with the added benefit of package management, which ensures that the versions of different libraries installed in your Python environment are compatible with each other. Detailed installation instructions are provided in [3].

Once you have installed Anaconda distribution successfully, we suggest that you downgrade Python version to 3.6 by executing the following set of commands from Anaconda Prompt, which can be accessed from the program/start menu and is also shown in Figure 2-1.

```
> conda update --all
> conda install python=3.6
```

```
(C:\Users\ayush\Anaconda3) C:\Users\ayush>conda install python=3.6
Fetching package metadata ...............
Solving package specifications: .

Package plan for installation in environment C:\Users\ayush\Anaconda3:

The following NEW packages will be INSTALLED:

    conda-package-handling: 1.6.0-py36h70ac491_2   conda-forge
    python_abi:             3.6-1_cp36m             conda-forge
    tqdm:                   4.47.0-pyh9f0ad1d_0     conda-forge

The following packages will be UPDATED:

    conda:            4.3.30-py36h7e176b0_0   --> 4.8.3-py36h9f0ad1d_1        conda-forge
    menuinst:         1.4.10-py36h42196fb_0   --> 1.4.16-py36_0               conda-forge
    pycosat:          0.6.2-py36hf17546d_1    --> 0.6.3-py36h68a101e_1004     conda-forge
    python:           3.6.3-h9e2ca53_1        --> 3.6.10-he025d50_1009_cpython conda-forge
    vc:               14-h2379b0c_2           --> 14.1-h869be7e_1             conda-forge
    vs2015_runtime:   14.0.25123-hd4c4e62_2   --> 14.16.27012-h30e32a0_2      conda-forge

The following packages will be SUPERSEDED by a higher-priority channel:

    conda-env:        2.6.0-h36134e3_1        --> 2.6.0-1                     conda-forge

Proceed ([y]/n)? y

vs2015_runtime 100% |###############################| Time: 0:00:00   10.04 MB/s
vc-14.1-h869be 100% |###############################| Time: 0:00:00    2.54 MB/s
python-3.6.10- 100% |###############################| Time: 0:00:02   10.70 MB/s
python_abi-3.6 100% |###############################| Time: 0:00:00    1.06 MB/s
tqdm-4.47.0-py 100% |###############################| Time: 0:00:00    5.03 MB/s
conda-package- 100% |###############################| Time: 0:00:00   10.53 MB/s
pycosat-0.6.3- 100% |###############################| Time: 0:00:00    4.41 MB/s
conda-4.8.3-py 100% |###############################| Time: 0:00:00   10.89 MB/s
```

Figure 2-1. *Installation of Python version 3.6 using Anaconda prompt*

These steps ensure that you have the Python version 3.6, which is required to install some of the machine learning libraries used later in Chapter 3. At this point, you should have all the Python libraries necessary to work through the rest of this chapter. Now, let's get familiar with the Jupyter Notebooks.

Jupyter Notebooks

A Jupyter Notebook, or simply *notebook*, provides an interactive way to write and execute Python code in a web-based interface. Notebooks also allow you to write markdowns, which makes it easier to share information about the code. If you want to make an attractive notebook, you can include images and other formatted text in the markdown cells. However, we are going to keep our focus on the Python code. To launch a new Jupyter Notebook for this chapter, navigate to the directory containing data and book source code in the command prompt or terminal, and use the following command:

```
> jupyter notebook
```

You should see an Internet browser window opening with a dashboard, as shown in Figure 2-2.

Figure 2-2. *Jupyter Notebook dashboard being displayed in a web browser*

Once you have successfully launched the notebook dashboard, locate and click the New button on the right-hand side of the dashboard, and select **Python 3** under Notebook:. You should see a notebook opening in a new browser tab, as shown in Figure 2-3.

Figure 2-3. *A new Jupyter Notebook being displayed in a web browser*

Congratulations! You have created your first Jupyter Notebook. In the web-based interface, you can see the menubar with different tabs, such as File, Edit, and so forth. Below the menubar, a toolbar provides the option to create new cells (denoted by the button with the + sign), which may be a code cell, a markdown cell, or other depending upon the selection made in the toolbar dropdown with the relevant options. While code cells contain regular Python code, markdown cells may contain a plain or formatted text description of the functionality being developed in the notebook. A cell can be executed by hitting Ctrl+Enter, or Alt+Enter. If you hit Ctrl+Enter, the code or markdown in the current cell is executed. However, if you hit Alt+Enter, a new cell is inserted below the current cell in addition to the execution of the code or markdown in the current cell.

Why Jupyter?

This process of executing Python code can be thought of as a hybrid between the Python interpreter and Python scripts. Some of the added advantages provided by Jupyter Notebook include a convenient means of sharing work via rich computational and data-driven narratives that mix code, figures, data, and text [4]. An additional advantage of this web-based environment is that it combines all the different pieces like code, text, images, equations, and visualization plots, together in one document. The notebooks are easily shareable and could be edited on the go since they do not have too much dependency on the associated system. This is the primary reason why Jupyter Notebooks are used in a teaching/academic environment.

These different approaches are usually selected based on the objectives of the code and the amount of experience the user has. For example, data scientists in the beginner phase usually start with Jupyter Notebooks since they provide an interactive environment without installing a separate IDE (integrated development environment). However, software developers typically prefer an IDE since they provide more advanced debugging procedures, Git integration, and an integrated terminal.

Now that you have learned a few basic things about notebooks, let's write and execute our first Python code.

Note   Code snippets used in the chapters are provided in the form of Jupyter Notebooks, which can be found in the book's code repository. These Jupyter Notebooks provide a comprehensive code base, which allows you to navigate through end-to-end problem workflows discussed in the book chapters.

Getting Started with Python

Now, in the code cell of the new notebook, let's type the following code, followed by Ctrl+Enter.

```
print("Hello World!")
```

You should see the following output.

```
Hello World!
```

Congratulations! You have written your first line of code in Python. In this one line of code, we asked Python interpreter to print the words "Hello World!" to the output console. In the subsequent sections of this chapter, we are going to take a more detailed look at Python programming.

You can create a new code cell in the already created notebook for each code snippet, and execute the code snippets as we go over different programming concepts in this chapter.

This chapter is focused on the basics of Python programming language, which is used in this book for the execution of machine learning algorithms. The list of topics covered in this chapter includes the following.

- Basics of the Python programming language

  - Executing code (three different methods): Python Interpreter, IPython, and .py scripts

  - Language structure and meaning: syntax and semantics of the programming language

  - Is everything an object in Python? Discussion of variables and operators in Python

  - How can we use numbers and perform operations in Python?

  - Why is Python considered a dynamic and strongly-typed language?

- Data structures: lists, tuples, sets, and dictionaries

- Control flow tools and iterators: if, else, for, and while statements

- Basic object-oriented programming: functions, methods, and classes

- Python libraries: How to use the vast library of modules and packages available in Python, along with other useful libraries?

  - NumPy: Scientific computing using arrays and matrices

  - pandas: Data analysis

  - Matplotlib: 2D plotting and visualization

The chapter uses an open source dataset from the oil and gas industry and provides a walk-through of basics of data analysis and visualization using Python libraries.

Python Basics

A primary difference between a compiled language like C, Fortran, or Java and an interpreted language like Python is the execution process. Python is a very flexible language that can be easily executed line by line and provides multiple different ways to execute code written in Python programming language. Three commonly used approaches are provided next.

Python Interpreter

Python Interpreter offers one of the most basic approaches to execute line-wise code. Once Python is installed on a system, running 'python' on the command prompt starts the process. The current version of Python is displayed and followed by >>>, indicating that Python code can be executed from this point onward (see Figure 2-4).

```
Command Prompt - python
Microsoft Windows [Version 10.0.18362.476]
(c) 2019 Microsoft Corporation. All rights reserved.

C:\Users\ayush>python
Python 3.7.1 (v3.7.1:260ec2c36a, Oct 20 2018, 14:57:15) [MSC v.1915 64 bit (AMD64)] on win32
Type "help", "copyright", "credits" or "license" for more information.
>>>
```

Figure 2-4. *Using Python Interpreter to run line-by-line code*

Python Scripts

More experienced programmers prefer to generate code in the form of a script with a .py extension. The entire code is executed using a Python terminal (see Figures 2-5 and 2-6).

```
Chapter02.py ×
Chapter02.py >
    # This is the code for Chapter 02 - Primer to Python Programming

    import math
    print('Chapter 02')
    a = 10
    print('Square Root of the variable is:', math.sqrt(a))
```

Figure 2-5. *Using an IDE to create a simple .py file*

```
ayush@Ayush-PC MINGW64 ~/OneDrive/ML_Book_Apress/oilandgas_ml/Chapter2/Code (master)
$ python Chapter02.py
Chapter 02
Square Root of the variable is: 3.1622776601683795
```

Figure 2-6. *Executing the code from .py file in an integrated terminal*

Language Structure and Meaning

This section briefly covers the syntax and semantics of Python programming language, which makes it easier to interpret, implement, and dynamically code for multiple programming paradigms. For an in-depth discussion on any of these concepts, you can refer to Guido van Rossum [5], who is the author of Python programming language. Python's simple and consistent design is based on the philosophy that "there should be one—and preferably only one—obvious way to do it," as mentioned in *The Zen of Python* [6].

The structure or form of a language is what constitutes the syntax for the programming language. It is usually comprised of three levels— words, phrases, and context—that define the rules in which a program is written. No information about the meaning or execution of a command is provided with the syntax. On the other hand, the semantics of a language is defined by the meaning behind the code and what it is supposed to provide as an outcome. Some of the rules and meanings behind the language are listed in this chapter.

Python is a case-sensitive language; for example, test_variable and Test_Variable are treated as two different objects. Unlike some other programming languages, Python does not require a command terminator. Example, use of a semicolon (;) after every statement is not required. However, if two different statements need to be executed in the same line of code, a semicolon can separate them, as shown next.

```python
test_var = 10; print (test_var)
```

Output

```
10
```

Whitespace limits the control of flow, but it does not affect the code. This is illustrated in the following code snippet.

```python
x1 = 10*2
x2 = 10 * 2
x3 = 10    *    2

print(x1)
print(x2)
print(x3)
```

Output

```
20
20
20
```

Indentation is very critical to define the start of a block of code and indicates a group of statements executed together. This is the case where whitespace matters. Python does not use curly brackets ({}) to define a code block, and hence indentation is what distinguishes a code block from the rest of the code. As an example, the following 'for' loop shows that the indented code printing 'Count' is printed thrice due to loop repetition, but the statement 'Outside the loop' is only printed once.

```
var_length = 3
for i in range(var_length):
    # indented code block which is a part of loop
    print('Count', i)

print ('Outside the loop') # not a part of loop
```

Output

```
Count 0
Count 1
Count 2
Outside the loop
```

The forward slash is used for assigning a path to a directory because Python follows UNIX-style path specifications. For example, the relative path for a folder in Windows, C:\Desktop\Python\Code, needs to be changed to path C:/Desktop/Python/Code.

Strings could be represented with single quotes (' '), double quotes (" "), or triple quotes (""" """). A convention is to include one-word strings with single quotes, a sentence with double quotes, and a paragraph or docstring with triple quotes.

```
word = 'Example'
sentence = "This is an example of a sentence"
para = '''This is a paragraph <br>
```

With a multiline statement enclosed in triple quotes (' ' '), comments are an integral part of the code to explain the thinking process of the programmer and help others to follow the intention of the code. In Python, single-line comments can be started with a pound sign (#), while multiline comments (comment block) can be enclosed in triple quotes.

For a deeper dive into code style and formatting, the PEP8 Style Guide [7] is a recommended reading which can help readers write code in a much more organized and consistent style.

Is Everything an Object in Python?

Python is an object-oriented programming language and treats everything as an object. With a loose definition of an *object* in Python, as compared to other programming languages, some objects do not necessarily need to have attributes nor methods. Neither are all the objects subclassable. The creator of the language, Guido van Rossum [8] wanted "all objects that could be named in the language (e.g., integers, strings, functions, classes, modules, methods, etc.) to have equal status," and hence classified them as "first-class everything." In other words, every object can be assigned to a variable, included as a member of a list, used as a key-value pair in a dictionary, or to be passed as an argument.

One of the advantages of Python being a dynamically typed and interpreted language is the simplistic nature in which the variables are defined. In this case, data types are checked, and the source code is translated during the execution of the program.

In Python, each variable name is bound only to an object. Usually, assignment operators carry out this operation. When a value is assigned to a variable, in many programming languages, it is thought to be stored in the form of a container or memory bucket [4]. In Python, this is better represented as a pointer instead of being a container. This is one of the reasons why there is no declaration of a *datatype*, which is very commonly found in other languages. The following code snippet provides an example of how variable names can point to objects of any type.

```
var1 = 5 # var1 is an integer
var2 = 'This is a String' # var2 is a string
var3 = [1,2,3] # var3 is a list
```

There are different categories of operators available for carrying out multiple operations on data. Some of them are listed next.

- Arithmetic operators (addition, subtraction, modulus, etc.)

- Bitwise operators (AND, OR, XOR, etc.)

- Assignment operators (+=, /=, etc.)

- Comparison operators (<=, !=, etc.) that output a boolean value (i.e., True or False)

- Boolean operations (and, or, not, etc.) that output a boolean value (i.e., True or False)

- Identity and membership operators (is, is not, not in, etc.)

Data Structures

Basic built-in datatypes that are commonly used in a program include integer, float, complex, boolean, string, and None datatypes. Apart from these basic ones, there are a few more data structures that are discussed in this section. These compound type data structures include lists, tuple, dictionary, and sets. A brief description, along with an example for each case, can help us understand their format and different ways in which they could be used.

It is recommended that the reader refers to Python documentation [9] for a detailed explanation of each data structure.

Lists

Enclosed within square brackets ([ ]), lists are an ordered and mutable array of elements, separated by a comma, which can include different data types like int, float, string, and so forth. It is possible to have duplicate elements in a list. A nested list is a "list of lists" and is often used for a two-dimensional matrix. Lists and their elements are accessed in different ways. The indexing starts at 0, which is the first element of the list.

Negative indexes can also access the elements of a list where the last element has an index of –1, the second last is –2, and so on. There are multiple methods available to access or modify elements of a list. Some of the examples include append(), pop(), count(), sort(), and many more. Since the lists are mutable, it is possible to modify/change the elements of a list as well as combine multiple lists.

```
# lists
list1 = [1,2,3] # simple list with integer values
list2 = [1,None,'Python',3.14] # list with mixed data types
# nested list
list3 = [ [1,2,3], [10,20,30], ['Example', 'Nested', 'List']]
list4 = [1,2,3,3,2,2,2,1,5,8,0] # list with repeated elements
```

```
print(list1[0])
print(list2[-2])
print(list3[2][1])
print(list4.count(2))
print(list3[2][1:3])
print(list1 + list2)
```

Output

```
1
Python
Nested
4
['Nested', 'List']
[1, 2, 3, 1, None, 'Python', 3.14]
```

Tuples

Enclosed within curved brackets (), tuples are very similar to a list, with a primary difference being their immutable nature (unchangeable). Tuples usually contain a heterogeneous sequence of elements accessed via unpacking or indexing. The elements are ordered and separated by a comma, just like a list. Elements are accessed in the same way as a list and the indexing starts with 0 as well. Similar to lists, negative indexes can access the elements of a tuple where the last element has an index of –1, the second last is –2, and so on. Some of the methods available for a tuple include count() and index(). Advantages of using a tuple instead of a list include fast execution compared to an equivalent list as well as situations where we do not want the data to be modified.

```
# tuples
tuple1 = (1,2,3) # simple tuple with integer values
tuple2 = (1,None,'Python',3.14) # tuple with mixed data types
```

```
# nested tuple
tuple3 = ([1,2,3], [10,20,30], ['Example', 'Nested', 'Tuple'])
tuple4 = (1,2,3,3,2,2,2,1,5,8,0) # repeated elements
print(tuple1[0])
print(tuple2[-2])
print(tuple3[2][1])
print(tuple4.count(2))
print(tuple3[2][1:3])
print(tuple1 + tuple2)
```

Output

```
1
Python
Nested
4
['Nested', 'Tuple']
(1, 2, 3, 1, None, 'Python', 3.14)
```

Dictionary

Enclosed within curly brackets ({}), a dictionary is another composite datatype which is also sometimes known as an associative array and consists of a collection of key-value pairs. It has certain similarities with a list since both datatypes are mutable, dynamic (increase or decrease size) and can be nested (dictionary within a dictionary). The primary difference lies in the way elements are accessed. Dictionary elements are accessed via keys instead of the index position. A dictionary can also be created using a built-in function, dict(), as shown in the following example. One of the restrictions for a dictionary is that a given "key" can only appear once. Some of the commonly used methods for a dictionary include clear(), items(), keys(), values(), and pop().

```python
# dictionary
dict1 = {
    'Texas': 'Permian Basin',
    'Colorado': ' DJ Basin',
    'North Dakota': 'Williston Basin'
}
# nested dictionary
dict2 = dict(
    USA = {'Exxon', 'Chevron', 'Conoco'},
    UK = 'BP',
    Revenue = [279.3, 158.9, 38.73, 303.7])
print(dict1)
print(dict2)
```

Output

```
{'Texas': 'Permian Basin', 'Colorado': ' DJ Basin', 'North
Dakota': 'Williston Basin'}
{'USA': {'Conoco', 'Chevron', 'Exxon'}, 'UK': 'BP', 'Revenue':
[279.3, 158.9, 38.73, 303.7]}
```

Sets

Enclosed within a curly brace ({}) and separated by a comma, sets are an unordered collection of items. Since the elements are not in any particular order, indexing has no meaning in sets. Sets can have elements of different data types but they have to be unique and must be immutable. However, the set itself can be changed and it is possible to add/remove items from it.

A big advantage of using the sets is the utility of being able to perform mathematical operations like union, intersection, symmetric difference and so forth. Although elements could be of different data types like int, float or string, a set cannot have mutable elements like list or dictionary as its element. An example is shown where the elements of a set are updated and the union of two sets are printed out. Some of the methods available for set are add(), clear(), discard(), intersection(), and pop(), and so forth.

```
# sets
set1 = {1,2,3} # simple set with integer values
set2 = set((1,None,'Python',3.14)) # list with mixed data types
# Gives an Error
# set3 = {[1,2,3], [10,20,30], ['Example', 'Nested', 'List']}
set4 = {1,2,3,3,2,2,2,1,5,8,0} # set with repeated elements

print(set1)
print(set2)
# print(set3)
print(set4)
```

Output

```
{1, 2, 3}
{3.14, 1, 'Python', None}
{0, 1, 2, 3, 5, 8}
```

Control Flow Tools and Iterators

Control flow is when a certain block of code is required to perform an operation in a repeated manner. Some of the simple conditional statements include if, elif, and else, which uses a boolean condition to execute a set of commands. The implementation of these concepts is illustrated using an example where we assume a simple case with only three types of rock permeability category. The use of indentations and semicolons (:) are a part of Python syntax.

```
# example of if-elif-else
perm = 0.001 # input in millidarcy

if (perm > 0.0001):
    print('This is a millidarcy case, seems to be a sandstone
    or limestone rock!')
elif(perm <= 0.0001 and perm>= 1E-6):
    print('This is a microdarcy case, seems to be tight sand
    rock!')
else:
    print('This is a nanodarcy case, seems to be shale rock!')
```

Output

```
This is a millidarcy case, seems to be a sandstone or limestone
rock!
```

On the other hand, there are two primary categories for loop control statements, which are either based on loop counts or controlled by an event. Examples of the loop controls include for, while, break, continue, and pass. A for statement belongs to the first category, where a loop is executed for every item in a list or a loop count. A while statement is executed while a particular condition is true. There are cases when the code needs to be forced out of the loop, and break is used in that situation. Continue is used when a remainder of a current loop needs to be skipped to another iteration.

```
Code for i in range(3):
    print('Entered the loop for count', i)
print('Loop Execution is complete!')
```

Output

```
Entered the loop for count 0
Entered the loop for count 1
```

```
Entered the loop for count 2
Loop Execution is complete!
Codewhile (perm<1E-6):
    print('Very Low Permeability typical for unconventional
    wells!')
else:
    print('Not an unconventional well!')
```

Output

```
Not an unconventional well!
```

Basic Object-Oriented Programming

While working on a project, often, situations are encountered where a block of code is required to be executed in a repeated manner with different arguments. The code is organized in much more reusable pieces, which are known as *functions*. In Python, they could be created by using def or lambda statements. An example of where a function can convert units is shown next. Every time we need to convert the units again, we can simply call the function and pass the new argument.

```
# function to convert units for mud weight
def unit_conversion(a):
    """

    This function is used to convert units of mud weight.

    Input: Mud Weight in kg/m3
    Output: Mud Weight in ppg or psi/ft or specific gravity

    """
    print('The entered mud weight in kg/m3 is:', a)
    mw_ppg = a * 0.0083454
```

```
    mw_psift = a * 0.000434
    sp_gravity = a * 0.001
    print('Mud Weight in ppg:', mw_ppg)
    print('Mud Weight in psi/ft:', mw_psift)
    print('Specific Gravity:', sp_gravity)
# enter the input for mud weight in kg/m3
mud_weight = 1250
unit_conversion(mud_weight)
```

Output

```
The entered mud weight in kg/m3 is: 1250
Mud Weight in ppg: 10.43175
Mud Weight in psi/ft: 0.5425
Specific Gravity: 1.25
```

When we are using only functions as blocks of statements within code, it is called a procedure-oriented way of programming. However, there is another way of writing more complicated and organized programs, commonly known as *object-oriented programming* (OOP). Data and functionality (i.e., properties and behavior) are wrapped inside a class and object, which are the two main aspects of OOP. A class creates user-defined data structures with information related to a particular characteristic. It can be thought of as a blueprint that can be filled out with different instances. Different objects are created within a class that contains attributes. An example of a simple class is shown next. The example illustrates how a "well" can be a class with its API number, Basin, and Production as attributes. For a more detailed description of classes, we recommended that you refer to [11].

```
class well():
    # class attribute
    well_type = 'oil well'
```

```python
    # initializer
    def __init__(self, api, basin, oil_prod, gas_prod):
        self.api = api
        self.basin = basin
        self.oil_prod = oil_prod # BBL
        self.gas_prod = gas_prod # MCF

    # instance method
    def boe_calc(self):
        boe = self.oil_prod + self.gas_prod/5.8
        return boe

# instantiating the class
well_1 = well(12345, 'Williston', 1000, 5000)

# calling a method of the class
print(well_1.boe_calc()) # boe production of well_1

well_2 = well(78910, 'Permian', 2000, 6000)
print(well_2.boe_calc()) # boe production of well_2
```

Output

```
1862.0689655172414
3034.4827586206898
# access the instance attributes
print('API for first well is:', well_1.api, 'and its in',
well_1.basin, 'Basin')
if well_1.well_type == 'oil well':
    print("{0} belongs to {1} category".format(well_1.api,
    well_1.well_type))
```

Output

```
API for first well is: 12345 and its in Williston Basin
12345 belongs to oil well category
```

Python Libraries

In the following section, some of the most popular libraries are briefly discussed. This includes pandas, NumPy, and Matplotlib. One of the other commonly used Python libraries is scikit-learn, which is used for machine learning. It is explained in later chapters.

pandas: Data Analysis Library

pandas is an open source library. It is derived from the term *panel data*, which is commonly used for data analysis in multidimensional structured datasets. Pandas can take multiple formats of a data file as input. Some of the examples include .csv, .xlsx, .json, .html, as well as many other commonly used formats. The data can then be arranged in the form of a dataframe consisting of rows and columns. Once data is converted into a dataframe, multiple different methods can be used for data exploration. It is also possible to obtain statistical information like mean, median, and standard deviation from the dataframe itself.

A dataset from the Volve field in the North Sea is used as an example in this chapter. Once the data is imported, a brief *exploratory data analysis* (EDA) is carried out by looking into the datatypes and descriptive statistics. The data is cleaned by dropping NA values, although a more thorough EDA process includes multiple steps, such as data imputation, outlier analysis, as well as preliminary feature selection. Another dataset with the same header information, but for a different well, is also imported, and the two dataframes are merged.

This dataset has the following petrophysical information available for analysis.

- Well Count: 2

- Well Name: 15/9-19 A and 15/9-19 BT2

- BVW (V/V): Bulk volume water

- CARB_FLAG (UNITLESS)

- COAL_FLAG (UNITLESS)

- KLOGH (mD)

- PHIF (V/V): Fracture porosity

- RHOFL (G/CM3): Bulk density fluid

- RHOMA (G/CM3): Bulk density matrix

- RW (OHMM): Water resistivity

- SAND_FLAG (UNITLESS)

- SW (V/V): Water saturation

- TEMP (DEGC): Temperature in degree C

- VSH (V/V): Shale volume

- LITHOTYPE: Rock characteristics

```python
import pandas as pd
#import the .csv file as a dataframe
data = pd.read_csv('data/15_9-19_A_CPI_LITHO.csv')

# data is imported as a table with rows and columns
data # entire dataset
```

Output

	COMMON_WELL_NAME	DEPTH (M)	BVW (V/V)	CARB_FLAG (UNITLESS)	COAL_FLAG (UNITLESS)	KLOGH (MD)	PHIF (V/V)	RHOFL (G/CM3)	RHOMA (G/CM3)	RW (OHMM)	SAND_FLAG (UNITLESS)	SW (V/V)	TEMP (DEGC)
0	15/9-19 A	3666.5916	0.111705	0.0	0.0	0.0003	0.199843	0.8	2.66	0.021643	0.0	0.5590	112.7249
1	15/9-19 A	3666.7440	0.112236	0.0	0.0	0.0000	0.218743	0.8	2.66	0.021642	0.0	0.5131	112.7285
2	15/9-19 A	3666.8964	0.114825	0.0	0.0	0.0000	0.242380	0.8	2.66	0.021642	0.0	0.4737	112.7321
3	15/9-19 A	3667.0488	0.114419	0.0	0.0	0.0000	0.226408	0.8	2.66	0.021641	0.0	0.5054	112.7358
4	15/9-19 A	3667.2012	0.115867	0.0	0.0	0.0000	0.216169	0.8	2.66	0.021640	0.0	0.5380	112.7394
...	...	...	...	...	...	...	...	...	...	...	...	...	...
3003	15/9-19 A	4124.2488	NaN	0.0	0.0	NaN	NaN	0.9	NaN	NaN	NaN	NaN	NaN
3004	15/9-19 A	4124.4012	NaN	0.0	0.0	NaN	NaN	0.9	NaN	NaN	NaN	NaN	NaN
3005	15/9-19 A	4124.5536	NaN	0.0	0.0	NaN	NaN	0.9	NaN	NaN	NaN	NaN	NaN
3006	15/9-19 A	4124.7060	NaN	0.0	0.0	NaN	NaN	0.9	NaN	NaN	NaN	NaN	NaN
3007	15/9-19 A	4124.8584	NaN	0.0	0.0	NaN	NaN	0.9	NaN	NaN	NaN	NaN	NaN

3008 rows × 15 columns

The following code snippet obtains the index and the different columns that constitute the dataframe. In order to access the elements/columns, either column names or numerical indexes can be used.

```
print(data.index)
print(data.columns)
```

Output

```
RangeIndex(start=0, stop=3008, step=1)
Index(['COMMON_WELL_NAME', 'DEPTH (M)', 'BVW (V/V)', 'CARB_FLAG
(UNITLESS)',
       'COAL_FLAG (UNITLESS)', 'KLOGH (MD)', 'PHIF (V/V)',
       'RHOFL (G/CM3)',
       'RHOMA (G/CM3)', 'RW (OHMM)', 'SAND_FLAG (UNITLESS)',
       'SW (V/V)',
       'TEMP (DEGC)', 'VSH (V/V)', 'LITHOTYPE'],
      dtype='object')
```

It is important to understand 1D (series) and 2D (dataframe) operations, which are defined next.

- **Series**: An array where an index is assigned to a value

- **Dataframe**: A 2D array where an index is assigned to multiple values and can include multiple columns

If you want to merge datasets from different sources, it is possible to combine them in multiple ways. A brief description of the different parameters used in the syntax is provided next.

```
pd.merge(left, right, how, on)
```

Here, the first two parameters indicate the first and second dataframe, which are required to be merged. These are often referred to as *left* and *right* dataframes. The parameter defines the type of merge to be performed, which could be one of the following.

- how = 'outer' allows you to take the union of datasets and append them together.

- how = 'inner' allows you to take the intersection of datasets and keep only the common values

- how = 'left'. All indexes of the left dataframe are included. If there are indexes that are not present in the 'left' dataframe, they will not be included in the merged dataset.

- how = 'right'. All indexes of the right dataframe are included. If there are indexes that are not present in the 'right' dataframe, they will not be included in the merged dataset.

The keyword 'on' decides if the dataframes are merged on a particular ID that is a unique identifier (for example, the API number in oil and gas wells).

```
# Merging data from another file
# Using the data from another well and merging the dataframes
data2 = pd.read_csv('data/15_9-19_BT2_CPI_LITHO.csv')
data_combined = pd.merge(data, data2, how="outer", on = None)
```

NumPy: Scientific Computing

This is a library in Python which is commonly used for scientific computing, using multidimensional array objects. Creating a NumPy array is a straightforward process. Examples of NumPy arrays creation using different methods are shown next.

```
# import the library
import numpy as np

array1 = np.array([1,2,3,4])
array2 = np.array([1,None,'Python',4.19])
array3 = np.array([[[(1,3.14,6),(1,2,3)], [(100,200,300),
(99,9.9,0.99)]]])
# evenly spaced values - step value
array4 = np.arange(5, 20, 4)
# evenly spaced values - number of samples
array5 = np.linspace(0,100,9)
array6 = np.random.randint(10, size=6)  # One-dimensional array
print(array1)
print(array2)
print(array3)
print(array4)
print(array5)
print(array6)
```

Output

```
[1 2 3 4]
[1 None 'Python' 4.19]
[[[  1.      3.14   6.   ]
  [  1.      2.     3.   ]]

 [[100.    200.    300.   ]
  [ 99.      9.9    0.99]]]
[ 5  9 13 17]
[  0.    12.5  25.    37.5  50.    62.5  75.    87.5 100.  ]
[6 7 3 6 1 1]
```

NumPy supports mathematical manipulations (addition, logarithm, etc.) and sorting of arrays. A simple example is shown in the following code snippet. To get a comprehensive understanding of the available NumPy operations, we encourage you to refer to the NumPy documentation [12].

```
# array mathematics - two different methods
print(array1 + array4)
print(np.add(array1, array4))
print(np.log(array1))

# sorting an array
array6.sort(axis=0)
print(array6)
```

Output

```
[ 6 11 16 21]
[ 6 11 16 21]
[0.         0.69314718 1.09861229 1.38629436]
[1 1 3 6 6 7]
print("Number of Dimensions: ", array3.ndim)
print("Array shape:", array3.shape)
print("Array size: ", array3.size)
```

Output

```
Number of Dimensions:  3
Array shape: (2, 2, 3)
Array size:  12
```

Accessing elements in NumPy arrays via indexing is very similar to accessing an element in the list. For a multidimensional array, elements can be accessed using a comma-separated tuple of indices. A *slicing operation* is also very similar. Some of the other more commonly used array manipulation routines are also available in NumPy, which are described in NumPy array manipulation documentation [13]. Some simple array manipulation examples are shown next.

```
i = np.append(array6, [25,26,27])
print(i)
concat = np.concatenate((array1, array2), axis=0)
print(concat)
mean_val = np.mean(array5)
print(mean_val)
```

Output

```
[ 1  1  3  6  6  7 25 26 27]
[1 2 3 4 1 None 'Python' 4.19]
50.0
```

Matplotlib: Python Data Visualization Library

Matplotlib is one of the most popular libraries for data visualization and was developed by John D. Hunter in 2003. There are multiple objects/components like axis, axes, and figure whose combination makes up a Matplotlib plot. There are different interfaces and backend rendering options available in Matplotlib. The library also offers different customization and designs to generate publication quality plots. Continuing with the same dataset, examples of a line plot, scatter plot, and a simple histogram are shown in Figures 2-7 through 2-9. Please refer to the provided Jupyter Notebook to access the code snippets used for generating these plots.

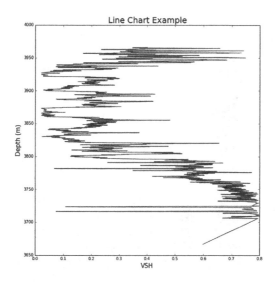

Figure 2-7. *Example of a line chart to observe shale volume (Vsh) with respect to depth*

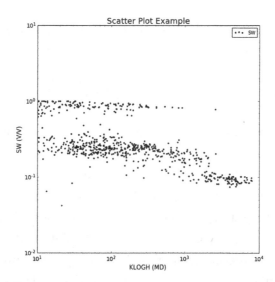

Figure 2-8. *Example of a scatter plot between K.Log(h) and water saturation*

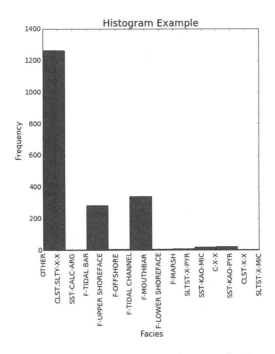

Figure 2-9. *Example of a histogram to observe the frequency of different facies in the dataset*

Summary

Python is a vast programming language, and it takes more than one book to cover all its programming concepts; however, in this chapter, we provided a brief introduction. We hope that you learned the basic data structures, control flows, and some of the standard libraries. This chapter provides a basic programming foundation necessary for this book. We strongly recommend that you refer to the original Python documentation and other resources mentioned in this chapter. These resources provide up-to-date descriptions and well-explained examples of ever-evolving Python programming concepts and libraries.

Acknowledgments

We thank Equinor AS, the former Volve license partners ExxonMobil Exploration and Production Norway AS, and Bayerngas (now Spirit Energy) for permission to use the Volve dataset, and to the many persons who have contributed to the dataset. Please visit data.equinor.com for more information about the Volve dataset and license terms of use.

References

[1] Python Software Foundation. [Online]. `https://www.python.org/`.

[2] Anaconda Inc. [Online]. `https://www.anaconda.com/distribution/#download-section`.

[3] Anaconda documentation. [Online]. `http://docs.anaconda.com/anaconda/install/`.

[4] J. VanDerPlas, *Python Data Science Handbook*, O'Reilly Media Inc., November 2016

[5] G. V. Rossum and F. L. Drake, Jr., "Python Tutorial: Release 3.2.3," June 2012

[6] [Online]. `https://www.python.org/dev/peps/pep-0020/`.

[7] [Online]. `https://www.python.org/dev/peps/pep-0008/`.

[8] "The History of Python," [Online]. `http://python-history.blogspot.com/2009/02/first-class-everything.html`.

[9] [Online]. `https://docs.python.org/3/tutorial/datastructures.html`.

[10] [Online]. `https://docs.python.org/3/tutorial/controlflow.html`.

[11] [Online]. `https://docs.python.org/3/tutorial/classes.html`.

[12] [Online]. `https://numpy.org/devdocs/reference/routines.math.html`.

[13] [Online]. `https://numpy.org/devdocs/reference/routines.array-manipulation.html`.

CHAPTER 3

Overview of Machine Learning and Deep Learning Concepts

In this chapter, we focus on exploring the realms of machine learning and deep learning algorithms. Datasets from the previous chapter are used in this chapter as well. In addition, we introduce some new public datasets to build machine learning and deep learning models. In the previous chapter, you learned the basic concepts of Python programming. In this chapter, we start using Python programming and related libraries to explore the data and doing data wrangling to clean it so that it can be used as an input for different machine learning models. The chapter is divided into seven broad sections, starting with the overview of machine learning.

This section is followed by some general concepts and best practices related to machine learning applications. The subsequent sections provide insight into exploratory data analysis, supervised learning, regression and classification algorithms, and unsupervised learning algorithms. Separate sections are dedicated to deep learning and hyperparameter optimization concepts and their application to selected oil and gas industry problems.

© Yogendra Narayan Pandey, Ayush Rastogi, Sribharath Kainkaryam,
Srimoyee Bhattacharya, and Luigi Saputelli 2020
Y. N. Pandey et al., *Machine Learning in the Oil and Gas Industry*,
https://doi.org/10.1007/978-1-4842-6094-4_3

Machine Learning

Machine learning is the field where computers are algorithmically programmed to learn and adapt themselves from experience to improve at a task when evaluated using a metric, rather than explicitly being told about the outcome [1]. This definition is explained with the simple example of a task in which the objective is to recognize handwritten digits, and the metric is the percentage of digits correctly classified using the database of human-labeled images of handwritten digits serving as experience. The "learning" part utilizes large amounts of data to enable the computer to understand underlying patterns.

With advancements in computational power and highly efficient machines, there has been a significant improvement in the efficiency of machine learning algorithms over the past few years. Some of the areas where machine learning has been successfully applied include image recognition, textual analysis (e.g., spam filtering), sentiment analysis or information extraction, video games, and robotics. With every passing day, more applications of machine learning are emerging.

Recent Advances

Recent advancements in the field of machine learning are seen in self-driving cars and autonomous vehicles, which have multiple sensors to gather real-time data at high frequency. The sensor data is processed by pre-trained machine learning algorithms, which provide "intelligence" to a self-driving car or an autonomous vehicle. Similar technology is also used in laser terrain mapping, path planning, and adaptive vision.

Another successful application of machine learning lies in automatic speech recognition, and language translators, which are commonly used in smart devices. Generative models built on the principle of generative adversarial networks (GAN) serve as another example of the most recent advances in machine learning, where reconstruction or generation of realistic human faces is possible using specialized deep learning models.

Machine Learning Categories

Machine learning algorithms can be divided into three broad categories.

- **Supervised learning:** In supervised learning, the desired output or response variable is known, and the machine learning algorithm provides a mapping between the input features and output variable. The two major subcategories of supervised learning are regression and classification problems, which are dictated by the type of output variable. When the output variable is continuous, it falls into the regression category.

    On the other hand, for classification problems, the output variable contains multiple classes or labels. In supervised learning, the model training process continues with the evaluation of error and making improvements until a desired level of accuracy is achieved.

- **Unsupervised learning:** In unsupervised learning, there is no explicit output variable, and the relationships are generated based on the data provided to the algorithm. Some of the algorithms belonging to this category can reveal hidden structures and relationships between the input features. Some of the examples of unsupervised learning include clustering, dimensionality reduction algorithms, and associative rule learning.

- **Reinforcement learning:** This category of algorithms is designed in such a way that there is a reward or penalty associated with the sequence of decisions made by the algorithm. A reward or penalty helps the algorithm learn the set of decisions it should make to achieve a defined objective. These algorithms are modeled using the Markov decision process (MDP).

  Reinforcement learning is often considered "semi-supervised" learning, but in an uncertain and potentially complex environment, the algorithm employs a trial and error approach to find solutions by being either penalized or rewarded for the actions it performs. Robotics for industrial automation is an example of reinforcement learning applications.

As discussed in Chapter 1, we keep the discussion limited to supervised and unsupervised learning algorithms in this chapter.

Model Training Considerations

With the availability of a vast number of algorithms in the data science toolkits, sometimes it becomes challenging to understand which algorithm must be used for solving a problem. Many algorithms may solve the same problem, and learn the relationship between the input features and output variable. However, the technique and learning process adopted by different algorithms may be significantly different.

An algorithm can outperform other algorithms when certain model parameters are changed. The model training process involves additional parameters called *hyperparameters*, which may include the following.

- The number of iterations for model training.

- The fraction of training data (batch size) that is used during each iteration.

- The fraction of estimation error propagated to change model parameters (learning rate).

The iterative process of tuning these hyperparameters for learning optimal model parameters is known as *hyperparameter optimization*. We discuss a simple technique for hyperparameter optimization in a later section of this chapter.

In addition to optimal model parameters, the selection of optimal number and type of input features can also improve the accuracy of a model. This makes feature engineering and feature selection very important aspects of the machine learning process.

It should be noted that we are providing a basic overview and a sample code implementation of selected machine learning algorithms in this section. A thorough understanding of each algorithm requires a comprehensive review of each technique, which is not possible in a single book chapter. We encourage you to explore and understand each algorithm and its associated model parameters and hyperparameters in depth before applying them to real-world problems. A list of parameters associated with each algorithm is found in the official documentation of the respective machine learning library.

Each machine learning algorithm has three primary components associated with it: *representation, optimization,* and *evaluation* [2]. The functions are represented in the form of numerical, symbolic, instance-based, or probabilistic graph models [3]. To improve the performance of the algorithm, optimization methods, such as gradient descent, dynamic programming, or evolutionary computation, are employed. The evaluation of these models is made through statistical metrics, which may include computations of precision, recall, and root-mean-squared error (RMSE).

Machine Learning Libraries

Scikit-learn is one of the most popular open source machine learning libraries in Python, which is built on other open source Python libraries, including NumPy, SciPy, and Matplotlib. Scikit-learn was originally developed in 2007 as a part of the Google Summer of Code project. It provides the implementation of a wide range of supervised and unsupervised learning algorithms with a Python programming interface.

TensorFlow is another open source software library developed by Google for numerical computations, which is highly popular for machine learning applications, such as shallow artificial neural networks and deep learning. TensorFlow allows the creation of dataflow graphs using *tensors,* which are multidimensional arrays through which computation occurs. The library is built to support parallel runs on multiple CPUs and GPUs and provides wrappers for many programming languages, such as Python, C++, and Java. In this book, we use the TensorFlow 2.x release.

Keras, which means *horn* in Greek, is an open source high-level neural network library, which is very user friendly, modular, and easy to work with Python. It does not handle low-level computation the way TensorFlow does but uses a backend engine to perform computations for the development of models. Keras uses the TensorFlow backend by default.

There are many other open source libraries, such as Theano, PyTorch, OpenCV, and Apache Spark ML, available for developing machine learning models. However, we chose TensorFlow as the library of choice due to wide community adoption, and completeness.

We do not provide the mathematical details of the algorithms. We stay focused on the hands-on implementation of machine learning algorithms using standard Python libraries. Please see the "Further Reading" section at the end of this book chapter to find some useful resources, which help improve understanding of the mathematical foundations of machine learning algorithms discussed in this chapter.

Note   The main Python libraries used in this chapter include
NumPy, SciPy, Matplotlib, scikit-learn, Keras, TensorFlow 2.x,
XGBoost, LightGBM, and Hyperopt. Some of these libraries should
be already present in the standard Anaconda installation. For the
missing libraries, please follow the installation instructions from the
help pages of individual libraries. The installation process may vary
significantly based on the operating system (e.g., Windows or Linux),
the hardware on the computer being used (with or without NVIDIA
GPUs), and so forth. We suggest that you find appropriate installers
based on your operating system and hardware, and install the
required libraries before proceeding with the rest of this chapter.

Machine Learning Pipeline

A machine learning pipeline is a series of processing elements, which
include processes, threads, routines, and functions, arranged in the form
of a flowchart to transform data from one representation to another. The
goal of creating a machine learning pipeline is to improve modularity
while focusing on repeatability and flexibility.

A production-ready machine learning project requires a carefully
architected machine learning pipeline. The model itself is only one of
the many components in the end-to-end pipeline. Some of the other
components of machine learning workflow include data collection,
data verification and preprocessing, feature extraction, model selection,
training and validation, prediction, evaluation, and deployment. Our focus
in this chapter is on building machine learning models.

General Concepts

In this section, we discuss general concepts, which apply to most of the machine learning model building exercises.

Data Preprocessing: Normalization and Standardization

For building a machine learning model, we need to preprocess the data so that it is adequate for the computational stability of the machine learning algorithm. An important step in data preprocessing involves rescaling the input features and output variable (for regression problems) to make their ranges consistent before feeding them as an input to the machine learning algorithm. To understand this concept, let's consider an example with different data ranges in input features or output variables. During the training process, we may have an input feature with a large range, such as cumulative production (range from 1000 bbls to 100,000 bbls). Whereas, another feature may have a very small range, such as porosity (range from 0.05 to 0.60).

During the iterative training process, the model parameters (e.g., coefficients or weights) associated with the input feature with large values differ significantly from those associated with the small values. Also, if the output variable has a very large range, the error metric, such as RMSE, can result in very large values. When this large error value is used to adjust the model parameters, which are already significantly large, it is possible that during the iterative update of model parameters, some values go beyond the numerical limit that the computer can handle. This may result in computational instability, eventually causing poor performance during the learning process. High sensitivity to input values can also cause severe generalization errors [4].

While working with data having multiple features in different units and ranges, it is important to preprocess the data for faster training, leading to a reduction in overfitting and more accurate predictions. As an example, this is observed during convergence in gradient descent optimization process [5].

There are multiple methods, such as standardization, normalization, and scaling available in the scikit-learn library, which can help process the data efficiently. The choice of preprocessing method depends on the algorithm and the parameter optimization process. A word of caution; these terms are often used interchangeably, depending on the customs within various fields.

Normalization

Normalization is a process of rescaling the data from the original range so that values lie in the range of 0 to 1 or –1 to 1. This method can be applied when the approximate upper and lower bounds of the data are known, data has few or no outliers present, and the data has a nearly uniform distribution. One of the utilities for scaling the data is `MinMaxScaler()` provided by the scikit-learn library, which uses the following equation.

$$X_{norm} = \frac{X - Xmin}{Xmax - Xmin}. \qquad (3.1)$$

Standardization (Z-Score Normalization)

When the features are rescaled to have the properties of a standard normal distribution (Gaussian distribution with a bell-shaped curve) with zero mean and unit standard deviation, the process is called *standardization*. This is important when a machine learning algorithm is used because it assumes that the input data is normally distributed. In this approach, rescaled values are calculated by using the following equation.

$$z = \frac{x - \mu}{\sigma}, \qquad (3.2)$$

where z is the standard z-score, *x* is the input data, μ is the sample mean of input data, and σ is the standard deviation from the mean. This method can be used in cases where the feature distribution does not contain extreme outliers.

Underfitting or Overfitting

When we train the machine learning models, one of the main objectives is to use it for generating predictions for the unseen data. An adequately trained model should demonstrate the same level of prediction accuracy for unseen data, which it had shown for the training data. However, in certain scenarios, the model may not be able to perform well on the unseen data. This may be caused due to underfitting or overfitting of the model.

- **Underfitting:** When the training process is stopped too early, or the model is insufficiently exposed to training data. In such a scenario, the model is not able to learn general patterns present in the training data. Poor prediction accuracy on both the training data and unseen data is an indicator of underfitting.

- **Overfitting:** One may choose to train a model using all the data available in a training dataset for a larger number of iterations so that it learns all the patterns present in the training data. As a result, the model can demonstrate very high accuracy on the training data. However, the model is so fine-tuned to the training data that it loses its ability to generalize beyond the training data it has seen. In such scenarios, the model exhibits poor prediction accuracy for unseen data, even though it was able to generate predictions for training data with high accuracy. This kind of scenario indicates model overfitting.

It is important to ensure that model does not suffer from either underfitting or overfitting. Machine learning model should be adequately trained so that it can generate predictions for unseen data with reasonable accuracy. In other terms, a machine learning model should be able to *generalize*.

Data Splitting: Training, Validation, and Test Datasets

As we just discussed, overfitting is one of the big challenges encountered during the training of machine learning models. This is a result of training the model on minute details/noise present in the training data. In such scenarios, the model begins to memorize the training data. This leads to good performance on training data, but the accuracy of the model significantly drops for unseen data. To account for overfitting and help the models become more *generalized*, splitting the dataset is considered a necessary step. As a rule of thumb, a dataset is often broken down into three sets of observations: training, validation, and test data [6], which are defined as follows.

- **Training data:** A subset of observations used for the model training process. This dataset is used by the algorithm to learn the parameters of the machine learning model.

- **Validation data:** A subset of observations, which is used for evaluating the prediction performance of the model during the training process. The trend observed in prediction errors using the validation data can help in identifying the number of iterations for adequate training. In a typical model training, the prediction error should keep on decreasing for both the training and validation dataset through the iterative training process. However, if we observe a point where prediction error continues to decrease for the training data but starts to increase for the validation data, training should be stopped. The point from where the validation error begins to increase indicates the beginning of overfitting.

- **Test data:** This is the subset of data that is only used to assess the performance of a fully trained model. The prediction accuracy on the test data is an indicator of the model's performance on unseen data encountered in a real-world scenario.

The following code snippet shows the application of a scikit-learn utility for splitting the dataset.

```
from sklearn.model_selection import train_test_split
X_train, X_test, y_train, y_test = train_test_split(X, y,
test_size=0.25, random_state=42)
```

Model Evaluation Metrics

To understand how well the model generalizes on an unseen dataset, and to evaluate its performance, it is important to use a quantitative metric based on which model performance can be evaluated.

In this section, we discuss some evaluation metrics to quantify the prediction accuracy of a model, based on which best-performing algorithm is selected. The metrics are also dependent on the type of problem. For quantitative outcomes in a regression model, some of the commonly used metrics include the coefficient of determination (R^2), and root-mean-squared error (RMSE). R^2 is calculated using the following equation.

$$R^2 = 1 - \frac{SS_{res}}{SS_{tot}}.$$ (3.3)

SS_{res} is the residual sum of squares, and SS_{tot} is the total sum of squares mathematically calculated using the following expressions.

$$SS_{res} = \Sigma\left(y_i - y_{reg}\right)^2,$$ (3.4)

$$SS_{tot} = \Sigma\left(y_i - \bar{y}\right)^2.$$ (3.5)

Here, y_i is the value of each data point, $\bar{y}$ is the mean value, and y_{reg} is the value predicted by the regression model.

On the other hand, the RMSE metric is calculated by using the following equation.

$$RMSE = \sqrt{\sum_{i=1}^{n} \frac{\left(\hat{y}_i - y_i\right)^2}{N}},$$ (3.6)

where $\hat{y}_i$ is the value predicted by the regression model, and N is the number of observations.

For classification problems, the model accuracy is defined as the ratio of the number of correct predictions to the total number of predictions. For a classification problem with N different classes, a *confusion matrix* or an *error matrix* is also used, which is an $N*N$ matrix containing N correct classifications on major diagonal and rest of the possible errors on off-diagonal entries. This matrix provides a visual approach to evaluate the performance of a machine learning model for classification problems. An illustration of a confusion matrix for binary classification (two classes) is shown in Figure 3-1.

	Predicted class	
	P	**N**
Actual class **P**	TP	FN
Actual class **N**	FP	TN

Figure 3-1. *An illustration of confusion matrix, where P = positive, N = negative, TP = true positive; FP = false positive; TN = true negative; FN = false negative*

Based on the matrix, some accuracy metrics can be determined, which include precision, recall, and F1-score.

Precision

Precision is defined as the ratio of the number of true positive (correctly predicted) samples for a given class to the total number of samples predicted to be of that class. A high precision output indicates that out of the samples predicted to belong to a given class by the model, most of the samples belonged to that class.

$$Precision = \frac{True\ Positive}{True\ Positive + False\ Positive}. \qquad (3.7)$$

Recall

Recall is defined as the ratio of the number of true positive samples to the total number of samples belonging to that class. A high recall indicates that very few samples belonging to a given class were misclassified as belonging to another class.

$$Recall = \frac{True\ Positive}{True\ Positive + False\ Negative}. \qquad (3.8)$$

F1-Score

The F1-score provides a harmonic mean of precision and recall. During model selection, the model with the highest F1-score is selected among multiple classification algorithms.

$$F1 = 2 \times \frac{Precision \times Recall}{Precision + Recall}. \qquad (3.9)$$

Reproducible Machine Learning

Machine learning algorithms provide a better predictive capability to a set of an industrially relevant problem than their conventional counterparts. However, for industrial adoption, it is important to ensure that the model training and predictions generated by the models are exactly reproducible. Sometimes, even while working on the same dataset and same algorithm, it may be difficult to reproduce the results and obtain the same performance.

These discrepancies can be attributed to several reasons, such as random initialization of weights, random shuffling of datasets, and stochastic nature of several machine learning algorithms. To address this issue, carefully setting the *random seed* as a part of machine learning algorithm implementation can take us one step closer to achieving reproducibility. Some examples of setting a random seed for an operating system environment variable, Python random module, NumPy, and TensorFlow, are shown in the following code snippet.

```
seed_value = 42
# Set Python random seed using environment variable
os.environ['PYTHONHASHSEED']=str(seed_value)
```

```
import random
# Set Python random seed using Python random module
random.seed(seed_value)
import numpy as np
np.random.seed(seed_value) # Set Numpy random seed
import tensorflow as tf
tf.random.set_seed(seed_value) # Set TensorFlow random seed
```

Bagging vs. Boosting

So far, we have discussed the selection of a single machine learning model to provide optimum and reproducible performance. However, in real-world scenarios, a single model based on an algorithm may not be able to provide the best solution. *Ensemble learning* was introduced to address this challenge; it happens when multiple machine learning models are trained using the same algorithm to provide a more powerful and robust predictive model.

The idea is to combine multiple weak learners to create a strong learning ensemble, which can provide higher accuracy than a single machine learning model. Bagging and boosting are two of the most common techniques in this category. They help with reducing noise, variance (i.e., avoiding overfitting), or bias (i.e., avoiding underfitting). Before defining these concepts, it is important to understand *bootstrapping*, which is a resampling technique used in statistics. During bootstrapping, several samples are drawn randomly from a provided dataset with a replacement. A bootstrap can quantify the uncertainty associated with a statistical learning method [7].

Bagging is the short form of *bootstrap aggregation*. While encountering high variance (overfitting) in some machine learning algorithms, such as decision trees, bagging is extremely useful where homogenous weak learners (i.e., models with poor performance) are combined by training them in parallel, and a prediction is provided through an averaging (regression) or voting (classification) process. This process leads to a more robust model, as well as reduces the chances of overfitting.

Boosting, on the other hand, learns from an ensemble of homogenous weak learners, but sequentially and adaptively. This approach tries to minimize the bias (underfitting) by training a model based on a previously trained model with poor performance, as opposed to reducing variance (overfitting) in bagging. In other words, bagging uses an equally-weighted average for generating predictions, while boosting uses a composite of multiple learners to provide a model with better performance.

Model Interpretability

In the field of statistics and machine learning, it is well known that an overly complex model might not always be the optimum solution to a problem. This is primarily because of the loss of model interpretability, as the model gets more complex. Interpretability is critical for many industries, where the model output is required to be explained, and model inference is the priority. For example, in a deep neural network with multiple layers and multiple neurons, it might be difficult to explain the effect of a specific input feature on the output.

This contrasts with a decision-tree based approach, or an even simpler linear regression model, where changes to a parameter are directly mapped to changes in the output variable. A significant amount of effort is devoted to better understand complex machine learning algorithms; however, it remains an area of ongoing research.

Exploratory Data Analysis (EDA)

In a real-world scenario, data is usually never available in a perfect format, ready to be fed to a machine learning model. The first step for a machine learning project involves carrying out an exploratory data analysis (EDA), which includes steps to detect inconsistencies, check assumptions, determine relationships among explanatory variables, as well as understand the relationship between features [8].

Exploratory data analysis is a critical process in understanding and preparing data for a machine learning model. Some of the steps in EDA include variable identification, univariate and bivariate analysis, treatment of missing values, removal of outliers, and transformation or creation of additional variables. The purpose of EDA lies in better understanding the data and its assumptions, improving its quality, and generating hypotheses of the analysis by finding clues about the tendencies of the data [9]. This involves using summary statistics and visualization techniques to answer questions about the provided data.

A brief exploratory analysis for the oil and gas data used in Chapter 2 is performed in this section. We continue the analysis from the previous chapter, where the data from the two wells were merged. The merged dataset is used as an input, and a dataframe is generated to perform EDA. The following code output shows the size of a merged dataframe.

```
Shape of the dataset (rows, columns): (5127, 15)
```

The dataframe consists of multiple NA values, as shown in Figure 3-2, which are required to be removed for cleaning the dataset. As observed from the figure, there are multiple columns with 236 missing values across multiple features, which need to be treated. In this case, instead of imputation (i.e., replacing missing values), the values are dropped since the NA occurs for the same set of data points. Once NA values are removed from the dataset, the new shape of the dataframe becomes the following.

```
Shape of the dataset (rows, columns): (4818, 15)
```

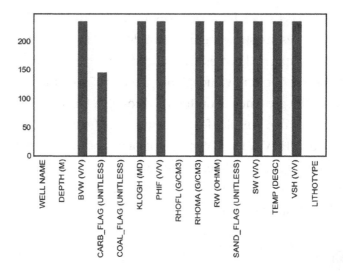

Figure 3-2. *Histogram of missing or NA values in the dataset*

The next step is to investigate the descriptive statistics of the features in the dataset. Figure 3-3 shows descriptive statistics of the dataframe.

	count	unique	top	freq	mean	std	min	25%	50%	75%	max
WELL NAME	5127	2	15/9-19 BT2	3162	NaN	NaN	NaN	NaN	NaN	NaN	NaN
DEPTH (M)	5127.00	NaN	NaN	NaN	3935.10	153.56	3666.59	3814.95	3912.57	4054.53	4249.83
BVW (V/V)	4891.00	NaN	NaN	NaN	0.10	0.06	0.00	0.06	0.09	0.12	0.25
CARB_FLAG (UNITLESS)	4980.00	NaN	NaN	NaN	0.03	0.18	0.00	0.00	0.00	0.00	1.00
COAL_FLAG (UNITLESS)	5127.00	NaN	NaN	NaN	0.01	0.07	0.00	0.00	0.00	0.00	1.00
KLOGH (MD)	4891.00	NaN	NaN	NaN	199.85	700.43	0.00	0.00	0.00	23.31	13705.31
PHIF (V/V)	4891.00	NaN	NaN	NaN	0.14	0.06	0.00	0.10	0.14	0.20	0.36
RHOFL (G/CM3)	5127.00	NaN	NaN	NaN	0.86	0.05	0.80	0.80	0.90	0.90	0.90
RHOMA (G/CM3)	4891.00	NaN	NaN	NaN	2.66	0.01	2.63	2.65	2.66	2.66	2.71
RW (OHMM)	4891.00	NaN	NaN	NaN	0.02	0.00	0.02	0.02	0.02	0.02	0.02
SAND_FLAG (UNITLESS)	4891.00	NaN	NaN	NaN	0.36	0.48	0.00	0.00	0.00	1.00	1.00
SW (V/V)	4891.00	NaN	NaN	NaN	0.76	0.28	0.04	0.56	0.89	1.00	1.00
TEMP (DEGC)	4891.00	NaN	NaN	NaN	117.87	2.97	112.72	115.43	117.60	119.84	124.09
VSH (V/V)	4891.00	NaN	NaN	NaN	0.64	0.62	0.02	0.21	0.57	0.98	17.12
LITHOTYPE	5127	20	OTHER	3010	NaN	NaN	NaN	NaN	NaN	NaN	NaN

Figure 3-3. *Descriptive statistics of the dataframe*

The next step is to look for the distribution of features. In Figure 3-4, some of the features are shown as an example. VSH (V/V) is one of the features, which displays skewness, possibly due to the occurrence of a wider range of less frequent values or occurrence of an outlier. A simple way to observe the presence of outliers is through a box plot. Figure 3-5 shows an example of a box plot for the same set of features. Again, looking into the feature VSH (V/V), you can see that a few points lie much farther outside the upper quartile range. In this section, the outlier analysis is not covered. However, you may take an additional filtering step to remove any existing outliers from the data.

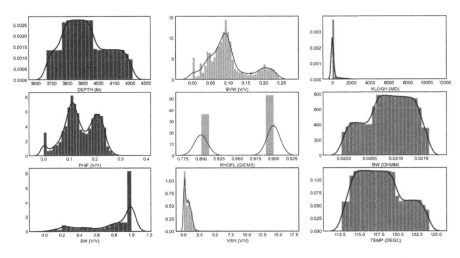

Figure 3-4. *Distribution of certain features in the dataset with multimodal and skewed behavior*

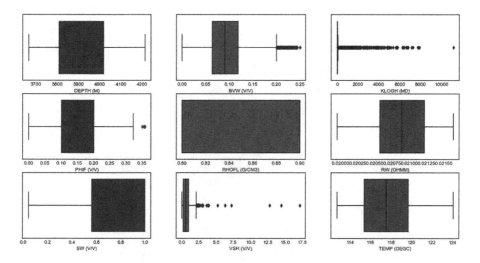

Figure 3-5. *Box plots of features in the dataset to identify the presence of outliers*

It is important to make sure there are no highly collinear variables in the dataset. A visual heatmap shown in Figure 3-6 observes the correlations between different features.

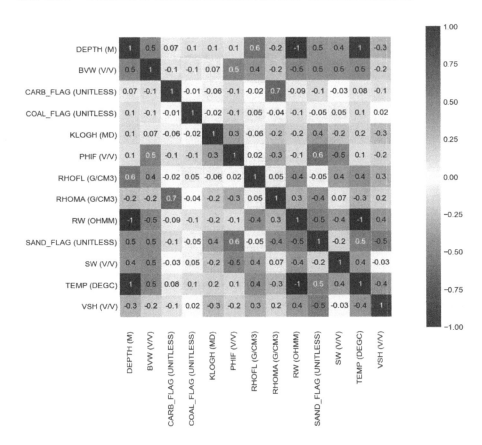

Figure 3-6. *Heatmap of correlation plot to understand any existing collinearity between variables*

Focusing on KLOGH (MD), which is the output variable in regression, as discussed later, no features are observed to be highly collinear except RHOMA (G/CM3) and CARB_FLAG (UNITLESS), which show a 70% collinearity. As a rule of thumb, it is advisable to take an additional step to understand a feature's importance and keep only the features of higher importance. However, feature engineering and feature selection are vast topics outside the scope of this chapter. We focus our discussion on building machine learning models.

Looking at the final list of variables in Figure 3-7, along with the number of observations, there is only one major categorical variable—LITHOTYPE, in the dataset. For classification problems, this is the output variable.

```
<class 'pandas.core.frame.DataFrame'>
Int64Index: 4818 entries, 0 to 4890
Data columns (total 15 columns):
WELL NAME               4818 non-null object
DEPTH (M)               4818 non-null float64
BVW (V/V)               4818 non-null float64
CARB_FLAG (UNITLESS)    4818 non-null float64
COAL_FLAG (UNITLESS)    4818 non-null float64
KLOGH (MD)              4818 non-null float64
PHIF (V/V)              4818 non-null float64
RHOFL (G/CM3)           4818 non-null float64
RHOMA (G/CM3)           4818 non-null float64
RW (OHMM)               4818 non-null float64
SAND_FLAG (UNITLESS)    4818 non-null float64
SW (V/V)                4818 non-null float64
TEMP (DEGC)             4818 non-null float64
VSH (V/V)               4818 non-null float64
LITHOTYPE               4818 non-null object
dtypes: float64(13), object(2)
memory usage: 762.2+ KB
```

Figure 3-7. *Datatype information about features in the dataset*

Supervised Learning

As we discussed at the beginning of this chapter, the problems where known input predicts the variable output falls under the supervised learning category. If the output is quantitative or continuous, it leads to solving a regression problem. Whereas, if the output is qualitative or discrete/categorical, it is a classification problem [10]. In other words, the objective is to learn a function f, which represents the systematic information that is provided by the input variables to predict or make inference about the response or dependent (output) variable [7].

In this section, we present code snippets of vanilla implementation of various regression and classification algorithms. The objective is to get you acquainted with different algorithms by providing a starting point. A later section in the chapter presents a hyperparameter tuning example, which can be adopted for any of the machine learning algorithms discussed in this chapter.

After providing a brief description and code snippets of different algorithms, a brief analysis is provided to understand the performance of each algorithm. However, it should be noted that these algorithm implementations are not optimized for performance; therefore, fine-tuned models using these algorithms may exhibit better performance than shown here using vanilla implementation.

Regression

Regression is a subtype of supervised learning problems, where the output is a quantitative or continuous variable. In the example of petrophysical data we have been working with, KLOGH is selected as the output variable. All other features are treated as independent input variables. Hence the objective of a regression problem is to find the optimal function f, which can make accurate predictions of KLOGH using all other variables as input features.

$$KLOGH = f(BVW, CARB\_FLAG, COAL\_FLAG, PHIF, RHOFL, RHOMA,$$

$$RW, SAND\_FLAG, SW, TEMP, VSH). \qquad (3.10)$$

Various machine learning algorithms are briefly discussed in this section without going into their mathematical details, or hyperparameter optimization. A minimal implementation of the algorithms is shown in this section, and we recommend that you delve deeper into each of these algorithms by following the suggestions provided in the "Further Reading" section at the end of this chapter.

The scikit-learn and TensorFlow machine learning libraries are used to implement the algorithms with Python. The end-to-end code for the examples shown in this chapter is provided in the Jupyter notebook available in the code repository of the book. The Jupyter notebook also contains some basic functions, which are defined to generate accuracy metrics, plot regression curves, and generate scatter plots of the predictions.

Multiple Linear Regression

Multiple linear regression, or simply multiple regression, is the simplest regression algorithm. In this algorithm, numeric coefficients are assigned to each input feature, and the output variable is predicted as the sum of the values obtained by multiplying feature value with the respective numeric coefficient. The training process of this algorithm tries to minimize the residual sum of squares between the observed and model-predicted output values. The following code snippet shows a sample implementation of this algorithm.

```
# Multiple Linear Regression
from sklearn.linear_model import LinearRegression
lin_reg = LinearRegression()
lin_reg.fit(X_train, y_train)  # training the algorithm
print(lin_reg.intercept_)  # intercept
print(lin_reg.coef_)  # coefficients
# Prediction on test data
y_pred_lin = lin_reg.predict(X_test)
```

Results

```
Mean Absolute Error: 283.89717843775026
Mean Squared Error: 245181.5892137209
Root Mean Squared Error: 495.15814566027376
R Squared: 0.1488228559603506
```

The regression plot for multiple linear regression is shown in Figure 3-8, along with a comparison of predicted values of KLOGH for the test dataset. As observed from the regression plot of observed and predicted values for test data, a linear model does not capture the nonlinear mapping between the input features and the output variable. It could be concluded that this algorithm is too simple to solve the problem and suffers from a strong bias (underfitting).

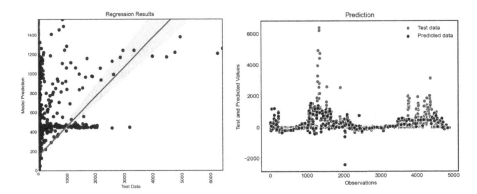

Figure 3-8. *Multiple linear regression results shown as regression plot and scatter plot*

Support Vector Regression

Support vector regression is based on a support vector machine. In this method, regression is performed on each point by allowing a margin for error. For each data point y, it is acceptable to get a predicted value in the range between $(y + \epsilon)$ and $(y - \epsilon)$. In Figure 3-9, the error margin ϵ is depicted by two supporting vectors and allows flexibility in model tuning.

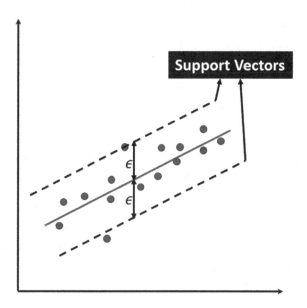

Figure 3-9. *Schematic depicting the support vectors with error margin* ϵ

In addition to the error margin, this algorithm takes advantage of a *kernel trick.* The kernel trick projects a nonlinear lower-dimensional space onto a higher-dimensional space. In the higher dimensional space, it is possible to represent the original nonlinear problem as a linear problem. Several kernels are available for support vector regression implementation in scikit-learn, which include `'linear'`, `'sigmoid'`, `'rbf'`, and `'polynomial'`.

A radial basis function (`'rbf'`) kernel, which is also called a *Gaussian kernel,* is selected in this example due to the presence of nonlinearity in the dataset. The following code snippet shows an example implementation.

```
# Support Vector Regression (SVR)
from sklearn.svm import SVR
svr_reg = SVR(kernel='rbf', C=100, gamma=0.1, epsilon=.1)
svr_reg.fit(X_train, y_train)
```

```
# Prediction on test data
y_pred_svr = svr_reg.predict(X_test)
```

Results

```
Mean Absolute Error: 82.06389051208197
Mean Squared Error: 112564.26261124818
Root Mean Squared Error: 335.5059799932755
R Squared: 0.6092197302512246
```

The regression plot and scatter plot based on the support vector regression results are shown in Figure 3-10. This algorithm performs significantly better than multiple linear regression. This improvement is attributed to the error margin and the kernel trick applied in the support vector regression.

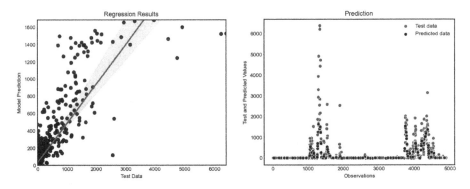

Figure 3-10. *Support vector regression results shown as regression plot and scatter plot*

Decision Tree Regression

Decision Trees are nonparametric tree-based flowcharts. This algorithm works by constructing a rule-based hierarchical tree structure built of multiple leaf nodes. A condition is imposed on a feature or a set of features at each leaf node. At each node of the decision tree, the samples in

102

training data are split into multiple child nodes depending on the imposed condition. The predicted values are calculated based on the average value of the samples in the terminal leaf nodes (see Figure 3-11).

During the training process, the algorithm learns the rules to split training data at each leaf node, so that the mean-squared-error (MSE) between the observed and predicted values is minimized. A decision tree is very commonly used for regression problems due to its highly interpretable nature. Figure 3-12 shows better predictions provided by the application of this algorithm when compared to the previous two algorithms. However, this algorithm suffers from the problem of overfitting. A decision tree regression code snippet is shown here.

```
# Decision Tree - Regression
from sklearn.tree import DecisionTreeRegressor
dt_reg = DecisionTreeRegressor(random_state=42)
dt_reg.fit(X_train, y_train)
# Prediction on test data
y_pred_dt = dt_reg.predict(X_test)
```

Results

```
Mean Absolute Error: 29.73391044401977
Mean Squared Error: 35097.83335297129
Root Mean Squared Error: 187.34415750957191
R Squared: 0.8781536833529513
```

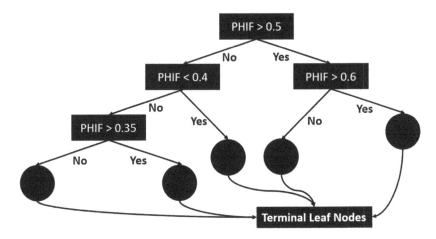

Figure 3-11. *Conceptual depiction of a decision tree, which splits data based on the value of PHIF*

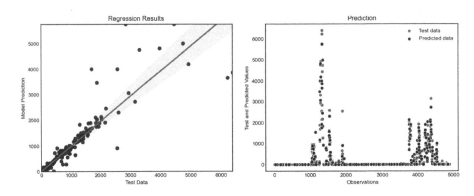

Figure 3-12. *Decision tree regression results shown as regression plot and scatter plot*

Random Forest Regression

A *random forest* is an algorithm based on an ensemble of decision trees. This algorithm applies bagging, which was discussed earlier in this chapter. A random forest algorithm applies bagging to both the input features and the training data. This means that for each decision tree grown in a random forest model, a subset of input features, and the training data is randomly selected to train that decision tree. The algorithm trains multiple decision trees in parallel.

The decision tree algorithm suffers from the problem of overfitting. This happens due to the use of a single tree, which almost memorizes the entire training data and fails to generalize. However, with the bagging of features and training data, this problem of overfitting is addressed in the random forest algorithm, which can provide a robust predictive model. The results obtained from the random forest regression show better prediction accuracy than the previously discussed algorithms. This observation can be verified by looking at the plots in Figure 3-13. A sample code implementation is shown here.

```
# Random Forest - Regression
from sklearn.ensemble import RandomForestRegressor
rf_reg = RandomForestRegressor(random_state=42)
rf_reg.fit(X_train, y_train)
# Prediction on test data
y_pred_rf = rf_reg.predict(X_test)
```

Results

```
Mean Absolute Error: 20.0011266166531
Mean Squared Error: 14744.082604216954
Root Mean Squared Error: 121.4252140381764
R Squared: 0.948814157854232
```

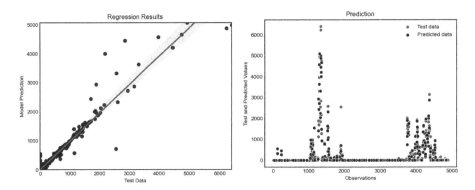

Figure 3-13. *Random forest regression results shown as regression plot and scatter plot*

XGBoost: eXtreme Gradient Boosting

XGBoost is among very powerful and popular machine learning algorithms, which provides high accuracy for large datasets due to its optimized implementation. This algorithm follows the boosting concept, where weak learners (decision trees) are grown sequentially. Each subsequent tree learns based on the prediction errors of the previous tree. A higher weightage is given to improving predictions at the data points, which exhibited higher MSE in the previous learner. An algorithm creates a composite model from weak learners, which provides robust performance.

The algorithm is designed to optimize performance at higher execution speed. The model itself is optimized using a gradient descent algorithm; as a result, it provides higher accuracy. Figure 3-14 shows a regression plot and a scatter plot drawn using the predictions from XGBoost regression. The following code snippet shows a sample implementation of the XGBoost regression algorithm.

```
# XGBoost - Regression
import xgboost as xgb
xgb_reg = xgb.XGBRegressor()
```

```
xgb_reg.fit(X_train, y_train)
# Prediction on test data
y_pred_xgb = xgb_reg.predict(X_test)
```

Results

```
Mean Absolute Error: 34.95929924279304
Mean Squared Error: 14853.490861224791
Root Mean Squared Error: 121.87489840498243
R Squared: 0.948434334034536
```

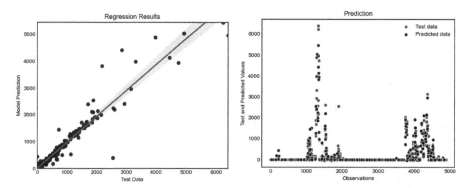

Figure 3-14. *XGBoost regression results shown as regression plot and scatter plot*

Artificial Neural Network

Artificial neural networks are built on the principle of decision-making capability of neurons in the human brain. These are the building blocks of more sophisticated deep learning networks. The neural network architecture consists of an input layer, one or more hidden layers, and an output layer. Each layer is made of neurons, where computations take place.

In a later section of this chapter, we discuss artificial neural networks and their application in deep learning in some more detail. In the following code example, a multilayer perceptron model is built, which uses the principle of backpropagation of error as a primary learning mechanism.

During the backpropagation of error, the estimated errors in model prediction are propagated back to adjust the neural network model parameters, and updated model parameters are calculated. This iterative training process continues until a certain condition for model accuracy, or training duration is satisfied. Figure 3-15 shows the plots based on prediction results obtained by the neural network model built using the following code.

```python
# Multi-Layer Perceptron - Regression
import tensorflow as tf
from tensorflow import keras
from tensorflow.keras import layers

def build_model():
    ann_reg = keras.Sequential([
        layers.Dense(32, activation="relu",
        input_shape=[len(X_train.keys())]),
        layers.Dense(32, activation="relu"),
        layers.Dense(1) ])

    optimizer = tf.keras.optimizers.RMSprop(0.001)
    ann_reg.compile(loss='mse', optimizer=optimizer,
    metrics=['mse'])
    return ann_reg
```

Results

```
Mean Absolute Error: 80.5055592187044
Mean Squared Error: 51228.191722683405
```

```
Root Mean Squared Error: 226.33645690140906
R Squared: 0.8221552194654382
```

The accuracy of the basic neural network model built here may not seem as good as some of the simpler models discussed earlier in this chapter. As the complexity of machine learning models increases, the task of training them adequately becomes more challenging. To train a *deep* neural network model adequately, we need to follow the process of hyperparameter optimization, which is discussed later in this chapter.

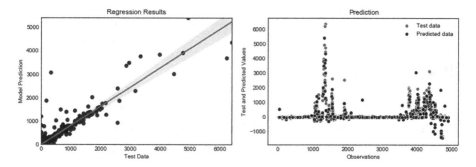

Figure 3-15. *Artificial neural network regression results shown as regression plot and scatter plot*

Comparison of the Regression Models

To compare the results of different machine learning models discussed earlier, we compare the RMSE score to evaluate their performance. Based on the results for the test dataset, random forest, and XGBoost regression algorithms seem to be among the best performers (see Table 3-1). However, we want to reiterate that these results are based on the simplest possible implementation of these algorithms, and without any of the parameters being tuned. The fine-tuned models for each of these algorithms show different accuracy than shown in this summary. However, this analysis establishes the efficacy of random forest and XGBoost algorithms in building quick prototypes for regression problems.

Table 3-1. *Regression Models and RMSE Scores*

Regression Model	RMSE
Multiple Linear Regression	495.16
Support Vector Regression	335.51
Decision Trees	187.34
Random Forest	121.42
XGBoost	121.87
Artificial Neural Network	226.34

Classification

Classification is a subcategory of supervised learning, where output variables are discrete or categorical. For the example discussed in this chapter, LITHOTYPE is the only categorical variable, and hence is used as the output variable. The objective of the classification algorithms is to find function f, which can map the features to predict different lithotype categories accurately.

$$LITHOTYPE = f(BVW, CARB\_FLAG, COAL\_FLAG, PHIF, RHOFL,$$

$$RHOMA, RW, SAND\_FLAG, SW, TEMP, VSH, KLOGH).   (3.11)$$

For the petrophysical dataset used in this chapter, there are multiple lithotype classes present in the data, as shown in Figure 3-16.

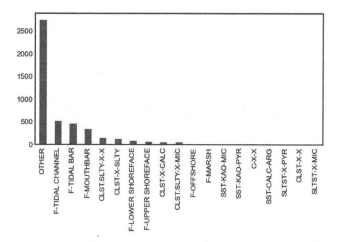

Figure 3-16. *Different lithotype classes available in the output variable*

For the sake of simplicity, only categories with at least 250 data points are kept, and the rest of them are filtered out. As observed from Figure 3-16, the category with the highest frequency is 'OTHER', which does not relay any useful information and hence is also discarded.

Figure 3-17 shows the top three categories, which are used in the classification model. A categorical output variable with only two classes is a binary classification problem. However, there are more than two classes in the output variable in the dataset that we are using. Therefore, we are solving a *multiclass* classification problem in this example.

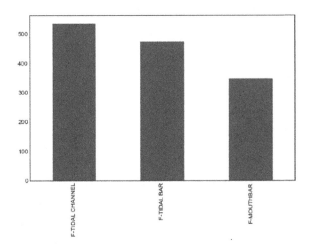

Figure 3-17. *Three major classes in the LITHOTYPE classification output variable*

Multinomial Logistic Regression

Logistic regression is a classification technique, which is used in binary classification problems. Binary classification is quite simple. If a sample doesn't belong to one class, it must belong to the other. Logistic regression uses the sigmoid function (see Figure 3-21). The sigmoid function changes from 0 to 1 for a very small change in the input value. This property of sigmoid function helps in classifying data with samples belonging to two distinct classes. For classification problems involving more than two classes, *multinomial logistic regression* is one of the basic multiclass classification algorithms. This algorithm uses a *softmax* function, which is expressed by the following equation.

$$f_{softmax}\left(z^i\right) = \frac{e^{z_i}}{\sum_{j=1}^{k}e^{z_j}}, \qquad (3.12)$$

where $k=$ number of classes in a multiclass classification problem, and i is the class for which softmax function is being computed corresponding to the input z^i. Softmax function also changes very rapidly from 0 to 1, but it has the capability of handling multiple classes. In simple terms, softmax function provides a probability that a sample belongs to either of the k classes. Multinomial logistic regression is also known as *softmax regression*. The following code snippet shows an example implementation of multiclass classification using this algorithm.

```
from sklearn.linear_model import LogisticRegression
clf_logreg = LogisticRegression()
clf_logreg.fit(X_train, y_train)
# Prediction on test data
y_pred_logreg = clf_logreg.predict(X_test)
# Accuracy Metrics
clf_metrics(y_test, y_pred_logreg)
```

Results

```
Classification Accuracy Score: 0.6480938416422287
Classification Report:
```

	precision	recall	f1-score	support
F-MOUTHBAR	0.40	0.05	0.08	87
F-TIDAL BAR	0.53	0.86	0.66	119
F-TIDAL CHANNEL	0.82	0.85	0.84	135
accuracy			0.65	341
macro avg	0.59	0.58	0.53	341
weighted avg	0.61	0.65	0.58	341

The results for the classification accuracy are shown in the form of a classification report, where precision, recall, and F1-score are shown as the quantitative indicators of the classifier's predictive capability. The algorithm predicts F-TIDAL CHANNEL class with higher precision but doesn't provide good precision for the classes with a lower population.

It should be noted that some of the classifiers don't work very well with the data, where one class has more samples than the others. In a *population imbalance* scenario, resampling the training data to make a population of classes equal may help in improving the precision and accuracy of the classification algorithm.

Support Vector Classifier

Earlier, we discussed support vectors and the kernel trick. The *support vector classifier* uses similar concepts. To perform classification, this algorithm constructs optimal hyperplanes that separate the samples belonging in different categories. An equivalent 2D representation can be thought of as a line dividing the cluster of points belonging to two separate categories. The algorithm uses a regularization parameter, which dictates the margin (recall ϵ) that separates the hyperplanes. The results indicate that this algorithm is unable to identify the classes with a lower population.

```
from sklearn.svm import SVC
clf_svc = SVC()
clf_svc.fit(X_train, y_train)
# Prediction on test data
y_pred_svc = clf_svc.predict(X_test)
# Accuracy Metrics
clf_metrics(y_test, y_pred_svc)
```

Results

```
Classification Accuracy Score: 0.6217008797653959
Classification Report:
```

	precision	recall	f1-score	support
F-MOUTHBAR	0.00	0.00	0.00	87
F-TIDAL BAR	0.50	0.87	0.64	119
F-TIDAL CHANNEL	0.80	0.81	0.80	135
accuracy			0.62	341
macro avg	0.43	0.56	0.48	341
weighted avg	0.49	0.62	0.54	341

Decision Tree Classifier

The decision tree classifier algorithm follows a similar approach, as discussed earlier for the decision tree regression. In a regression algorithm, the predictions were generated by averaging the values in the terminal leaf node, to which a sample belongs. In a classification problem, all samples belonging to a terminal tree node are classified as one of the categories.

During the model training, the algorithm tries to minimize entropy (randomness) and maximize the information gain. An adequately trained tree with no randomness has all samples belonging to only one of the sample classes in any given terminal leaf node. The results show that despite population imbalance, this algorithm provides high precision and accuracy across the lithotype categories.

```
from sklearn.tree import DecisionTreeClassifier
clf_dt = DecisionTreeClassifier()
clf_dt.fit(X_train, y_train)
# Prediction on test data
y_pred_dt = clf_dt.predict(X_test)
# Accuracy Metrics
clf_metrics(y_test, y_pred_dt)
```

Results

```
Classification Accuracy Score: 0.9853372434017595
Classification Report:
                precision    recall  f1-score   support

   F-MOUTHBAR        0.98      0.98      0.98        87
  F-TIDAL BAR        0.99      0.97      0.98       119
F-TIDAL CHANNEL      0.99      1.00      0.99       135

    accuracy                            0.99       341
   macro avg         0.98      0.98      0.98       341
weighted avg         0.99      0.99      0.99       341
```

Random Forest Classifier

Random forest classifier works on the exactly similar concept, which we discussed for the random forest regression. The only difference is that in a classification algorithm, a forest of decision tree classifiers is grown, and the final prediction is generated based on the *voting* instead of averaging.

In the voting approach, the class that appears with the highest frequency in the ensemble predictions is selected as the final output. The algorithm also provides the flexibility of predicting probabilities of a sample belonging to any of the sample classes. The selection of one of these two output methods depends on the problem at hand. A sample code implementation and classification results show high precision and accuracy of predictions provided by this algorithm.

```
from sklearn.ensemble import RandomForestClassifier
clf_rf = RandomForestClassifier()
clf_rf.fit(X_train, y_train)
# Prediction on test data
y_pred_rf = clf_rf.predict(X_test)
```

```
# Accuracy Metrics
clf_metrics(y_test, y_pred_rf)
```

Results

```
Classification Accuracy Score: 0.9882697947214076
Classification Report:
                  precision    recall  f1-score   support

      F-MOUTHBAR       0.99      0.98      0.98        87
     F-TIDAL BAR       0.99      0.98      0.99       119
 F-TIDAL CHANNEL       0.99      1.00      0.99       135

        accuracy                          0.99       341
       macro avg       0.99      0.99      0.99       341
    weighted avg       0.99      0.99      0.99       341
```

k-Nearest Neighbors (*k*-NN)

k-nearest neighbors (*k*-NN) is a nonparametric algorithm based on the
similarity measure (e.g., distance functions, such as Euclidian distance
between two samples, or Hamming distance between the binary vectors).
The algorithm uses the similarity measure to find *k* nearest neighboring
points in the sample data, and based on the most frequently occurring
class among these *k* neighbors, predicts the output category.

One of the input parameters for this algorithm is the number of
neighbors that are searched in the proximity to a sample point with
an unknown category. The sample implementation and results of this
algorithm demonstrate a reasonable performance. However, it does not
perform as well as the previously discussed tree-based algorithms.

```
from sklearn.neighbors import KNeighborsClassifier
clf_knn = KNeighborsClassifier()
clf_knn.fit(X_train, y_train)
```

```
# Prediction on test data
y_pred_knn = clf_knn.predict(X_test)
# Accuracy Metrics
clf_metrics(y_test, y_pred_knn)
```

Results

```
Classification Accuracy Score: 0.8592375366568915
Classification Report:
```

	precision	recall	f1-score	support
F-MOUTHBAR	0.83	0.75	0.79	87
F-TIDAL BAR	0.86	0.85	0.86	119
F-TIDAL CHANNEL	0.87	0.94	0.90	135
accuracy			0.86	341
macro avg	0.86	0.85	0.85	341
weighted avg	0.86	0.86	0.86	341

Gaussian Naive Bayes Classification

The Gaussian Naive Bayes algorithm is based on *Bayes' theorem* with an assumption that multiple input features independently contribute to the probability of a sample belonging to a particular class. This algorithm assumes a Gaussian or normal distribution of the continuous features associated with each of the classes.

By using the principle of maximum likelihood, the algorithm can predict the probability that a sample point belongs to a particular class. This algorithm also falls under the category of basic classification algorithms but performs better than other basic classification algorithms, such as multinomial logistic regression. A sample implementation and classification results of this algorithm are shown in the following code snippet.

```
from sklearn.naive_bayes import GaussianNB
clf_gnb = GaussianNB()
clf_gnb.fit(X_train, y_train)
# Prediction on test data
y_pred_gnb = clf_gnb.predict(X_test)
# Accuracy Metrics
clf_metrics(y_test, y_pred_gnb)
```

Results

Classification Accuracy Score: 0.6099706744868035
Classification Report:

	precision	recall	f1-score	support
F-MOUTHBAR	0.40	0.54	0.46	87
F-TIDAL BAR	0.61	0.34	0.44	119
F-TIDAL CHANNEL	0.77	0.89	0.82	135
accuracy			0.61	341
macro avg	0.59	0.59	0.57	341
weighted avg	0.62	0.61	0.60	341

Linear Discriminant Analysis

Linear discriminant analysis (LDA) also works under the assumption of Gaussian distribution of input features. Also, the algorithm assumes that each feature has the same variance. In this algorithm, input features are transformed, such that there is minimal overlap between the values of the input features for two distinct classes. This is achieved by performing the transformation, which maximizes the distance between the statistical means of the features for any two distinct classes in the sample. The algorithm provides the probability of a sample point belonging to a particular class.

LDA is a simple algorithm, which can perform reasonably for a small dataset. However, if the training data contains input features with overlapping statistical means, the algorithm is not able to distinguish samples belonging to different classes. The following code snippet and results demonstrate the better performance of LDA on the multiclass classification problem when compared to other basic algorithms (i.e., Gaussian Naïve Bayes, and multinomial logistic classifier).

```
from sklearn.discriminant_analysis import
LinearDiscriminantAnalysis
clf_lda = LinearDiscriminantAnalysis()
clf_lda.fit(X_train, y_train)
# Prediction on test data
y_pred_lda = clf_lda.predict(X_test)
# Accuracy Metrics
clf_metrics(y_test, y_pred_lda)
```

Results

```
Classification Accuracy Score: 0.7800586510263929
Classification Report:
```

	precision	recall	f1-score	support
F-MOUTHBAR	0.73	0.56	0.64	87
F-TIDAL BAR	0.71	0.82	0.76	119
F-TIDAL CHANNEL	0.87	0.89	0.88	135
accuracy			0.78	341
macro avg	0.77	0.76	0.76	341
weighted avg	0.78	0.78	0.78	341

Comparison of Classification Models

For the classification models that we built in this section, F1-scores are used as a metric to compare their performance. Based on the results for the test dataset, decision tree, and random forest classifiers are among the best performing classification models, as shown in Table 3-2. Further tuning of these models is expected to change these accuracy results.

Table 3-2. *Classification Models and F1-Scores*

Classification Model	F1-Score
Multinomial Logistic Regression	0.648
Support Vector Classifier	0.621
Decision Tree Classifier	0.985
Random Forest Classifier	0.988
k-NN	0.859
Gaussian Naive Bayes	0.609
Linear Discriminant Analysis	0.780

Unsupervised Learning

So far, we discussed the problems related to supervised learning, where the machine learning algorithm tries to map input features in the training data to the corresponding output variable. In unsupervised learning problems, there is no associated output variable to be mapped to the input features. Instead, unsupervised machine learning algorithms work by finding hidden relationships and/or patterns in the training data. Two of the broad categories of unsupervised learning problems include clustering and dimensionality reduction. We discuss both categories of algorithms and show example implementations.

Clustering

In this category of unsupervised learning, algorithms try to find clusters with similar characteristics in the sample data. Clustering algorithms work very well on the sample data, where relatively distinct groups of the sample population are present. Once the clustering algorithm is trained, any new observation is predicted to belong to one of the clusters, which the algorithm had identified based on the original sample data.

K-Means Clustering

The k-means clustering algorithm is used for partitioning a dataset with n observations into k distinct clusters centered at a centroid. Each observation in the training data belongs to a particular cluster. A cluster or aggregated group of points have certain similar characteristics embedded into them. For the petrophysical dataset being used in this chapter, it is important to find the number of clusters for which an elbow plot is generated, as shown in Figure 3-18. The elbow plot shows the variation of *inertia* (the sum of squared distances of points from their cluster centroid) as a function of the number of clusters.

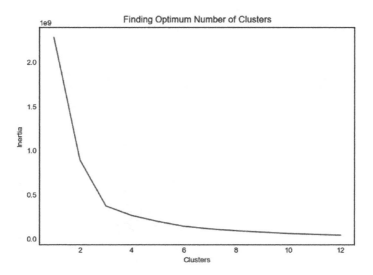

Figure 3-18. *Elbow plot indicating optimum number of clusters based on inertia value*

Looking at the elbow plot, for $k = 3$ clusters, a distinct kink or elbow is present. Using this optimum number of $k = 3$ clusters for the dataset, a scatter plot generated between KLOGH and PHIF (V/V) shows distinct boundaries between the three different clusters (see Figure 3-19). Physically, these clusters may depict logical grouping of the subsurface rocks, similar to facies or lithotypes.

```
import sklearn.cluster as cluster
import seaborn as sns
km3 = cluster.KMeans(n_clusters=3, init='k-means++', max_
iter=300, n_init=10, random_state=42).fit(X)
X['Labels'] = km3.labels_
X['Labels'].value_counts()
sns.scatterplot(X['PHIF (V/V)'], X['KLOGH (MD)'],
hue=X['Labels'], palette=sns.color_palette('hls', X['Labels'].
nunique()))
plt.title('KMeans with 3 Clusters')
```

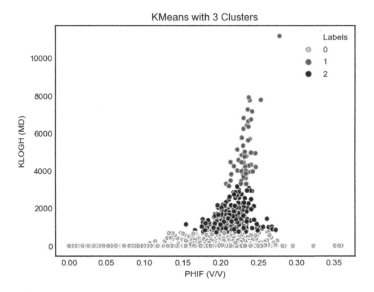

Figure 3-19. *Scatter plot between KLOGH and PHIF showing the boundaries between different clusters generated using the k-means clustering algorithm*

Dimensionality Reduction

Dimensionality reduction is an abstract concept, which is very useful for analyzing the datasets with an extremely high number of features. Dimensionality reduction is often performed to obtain better input features for machine learning algorithms. It improves computational efficiency without sacrificing much on the prediction capability of the models. Dimensionality reduction also removes the collinearity, which may exist between different features in the original data. Principal component analysis is an algorithm in this category.

Principal Component Analysis

Using the orthogonal transformation of features, *principal component analysis* (PCA) provides a statistical approach to reduce the number of dimensions in the input data by converting multiple linearly dependent features into linearly independent variables. These linearly independent variables are called *principal components*, which are orthogonal to each other.

The first principal component explains maximum variance in the dataset, which further reduces with each succeeding component. This method comes with a caveat that the principal components often turn out to be less interpretable than the original features. It is to be noted that PCA is sensitive to scaling, and hence it is important to scale the features using methods discussed earlier in this chapter.

Using the same dataset that we used earlier in this chapter, the following implementation shows that the first four principal components can explain approximately 86% of the sample variance.

```
from sklearn.decomposition import PCA
pca = PCA()
X_train = pca.fit_transform(X_train)
X_test = pca.transform(X_test)
explained_variance = pca.explained_variance_ratio_
print(explained_variance)
```

Results

```
Explained Variance
[4.32302153e-01 2.49297862e-01 1.06117586e-01 7.90285832e-02
 4.83399458e-02 3.41638092e-02 2.32342217e-02 1.67907297e-02
 6.98516043e-03 2.81954933e-03 9.13677877e-04 6.64565466e-06
 7.64286168e-08]
```

After the calculation of principal components, we use random forest classification to understand how reducing the number of input features following PCA affects the accuracy of machine learning models. Here, we compare the accuracy of predictions with all 12 features, with the prediction generated using the first four principal components. From the following code implementation results, reducing the number of input features to the first four principal components does not reduce prediction accuracy significantly.

```
# Random Forest with entire data
classifier = RandomForestClassifier(random_state=42)
classifier.fit(X_train, y_train)
# Predicting the Test set results
y_pred = classifier.predict(X_test)
print(confusion_matrix(y_test, y_pred))
print(accuracy_score(y_test, y_pred))

Accuracy with all features: 0.9377289377289377

# Random Forest with only 4 components
classifier_4comp = RandomForestClassifier(random_state=42)
classifier_4comp.fit(X_train_4comp, y_train)
# Predicting the Test set results
y_pred_4comp= classifier_4comp.predict(X_test_4comp)
print(confusion_matrix(y_test, y_pred_4comp))
print(accuracy_score(y_test, y_pred_4comp))

Accuracy with only first 4 principal components:
0.8974358974358975
```

Deep Learning

We have discussed supervised and unsupervised machine learning algorithms for solving problems related to classification, regression, dimensionality reduction, and clustering. While discussing the algorithms for regression, we briefly touched on artificial neural networks. Neural computation is at the heart of deep learning algorithms. Deep learning algorithms attempt to create deep hierarchical models, which can perform complex tasks, such as identifying one object from another, understanding human speech, or creating new paintings based on multiple input images.

In this section, we discuss the following deep learning algorithms.

- Multilayer perceptrons (MLP)

- Convolutional neural networks (CNN)

- Recurrent neural networks (RNN)

- Long short-term memory (LSTM)

Multilayer Perceptron (MLP)

A multilayer perceptron (MLP) is the most basic kind of deep learning algorithm. Neurons are the basic building blocks of deep neural networks, including MLP. Figure 3-20 shows a schematic representation of a neuron. Neurons compute a weighted sum of the input values by using the weights associated with the input and then apply a nonlinear *activation function* (*f*) to the weighted sum. Nonlinear transformations performed by the neurons provide MLP and other deep neural networks capability of modeling highly nonlinear processes, which occur in a lot of natural occurrences, and industry problems. Neurons are also known as *neural network nodes*.

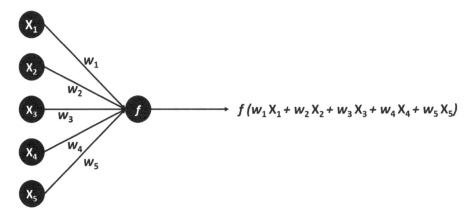

Figure 3-20. *Schematic depiction of computations in a neuron of artificial neural network*

The neurons may use several activation functions to perform nonlinear transformations. The sigmoid function and hyperbolic tangent function are among the classically used activation functions. Some of the more advanced activation functions include the *rectified linear unit* (ReLU), leaky ReLU, and exponential LU. A comparison of functional forms and shapes of these traditional and modern nonlinear activation functions [11] is shown in Figure 3-21. These functions are easy to differentiate (continuous and differentiable). This quality makes these functions computationally efficient during the gradient calculations, which are an important part of training deep neural networks.

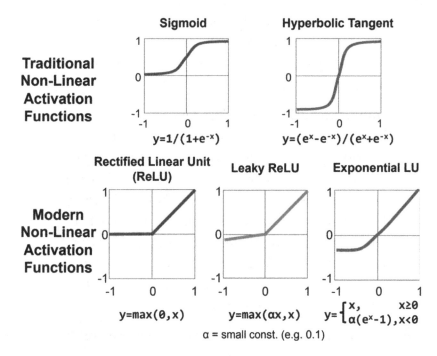

Figure 3-21. *Comparison of traditional and modern activation functions [11]*

Figure 3-22 shows a schematic depiction of an MLP. The following components constitute this kind of deep neural network.

- **Input layer:** This layer contains nodes, which facilitate the intake of input features to the neural network, apply nonlinear transformations.

- **Hidden layers:** These layers contain neurons or nodes with a nonlinear activation function. The hidden layer is represented by a weight matrix (W), and a bias vector (b). The dimensions of matrix W and vector b depend on the number of nodes in the layers. For example, the input layer, which connects with m input (feature matrix X) and broadcasts values to n nodes in the first hidden layer, has a weight matrix with dimension $m \times n$. Also, the bias vector b has n elements. The output from this layer is $f(WX + b)$, where f is the activation function.

- **Output layer:** The output layer connects with the last hidden layer, and ensures that the neural network provides output consistent with the problem formulation. For example, if the neural network is being trained for classifying a dataset with k distinct classes, the output layer computes k distinct values, and then apply a softmax function to each of these k values. In the described scenario, each of the elements in the output corresponds to the probability of the sample belonging to one of the k classes.

- **Optimizer:** The optimizer facilitates updates of weight matrix W and bias vector b for the neural network layers by using the backpropagation of errors.

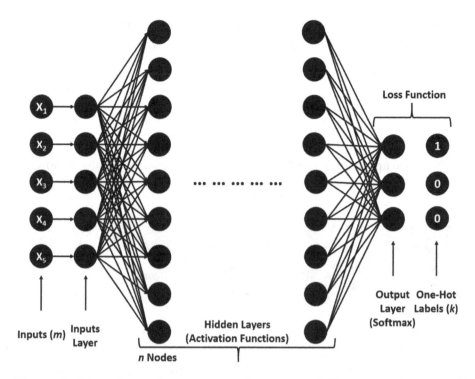

Figure 3-22. *Schematic representation of a multilayer perceptron*

The optimizer uses a loss function to evaluate the accuracy of the predicted values. In regression problems, this loss function can be RMSE. However, in classification problems, a relatively complex loss function may be required. In a classification problem, the output layer provides k probability values (y). If an object belongs to only one of the object classes, which is the case for most of the real-world classification tasks, one of these k values should be close to 1 for the optimal performance of the classifier. Labels are one-hot encoded to convert class labels to probabilities. For example, in a problem with $k=3$ classes, classes 1, 2, and 3 are represented as one-hot vectors (p) [1, 0, 0], [0, 1, 0], and [0, 0, 1], respectively. A loss function, such as categorical cross-entropy, uses these values to provide quantitative estimation of error, which is useful in iteratively updating neural network model

131

parameters (i.e., weights and biases). By using the output vector (y) and one-hot encoded label vector (p), the categorical cross-entropy can be computed using the following equation.

$$crossentropy = \sum_{i=1}^{k} - p_i \log(y_i). \qquad (3.13)$$

Next, we apply these concepts to classify images in the dataset shown in Figure 3-23.

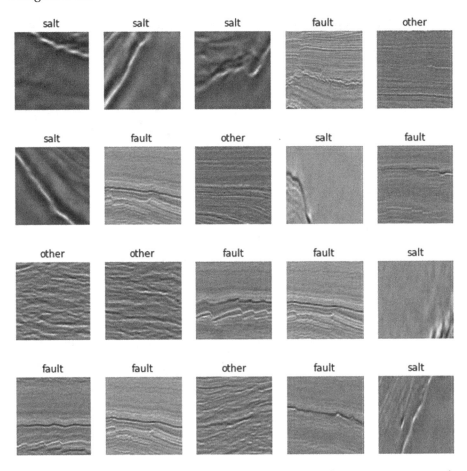

Figure 3-23. *Some sample images from the SFO (Salt/Fault/Others) dataset*

There are some standard image datasets available in the machine learning community, which are commonly used for evaluating classification models. These datasets include collections of handwritten digits (MNIST), objects encountered in routine life (CIFAR10), and so forth. For this chapter, we created a dataset using the seismic surveys available in the public domain and on Kaggle. This dataset (nicknamed *SFO dataset*) contains 1,500 seismic images belonging to three categories: salt, fault, and others. We used MLP to classify images in this dataset. The following code snippet shows an implementation of the MLP model using the TensorFlow library.

```
# Define a Deep Neural Network model
reg_param = 0.02 # Regularization parameter for L2
regularization
model = Sequential()
model.add(Dense(1024, activation="relu", kernel_
initializer="he_normal", kernel_regularizer=l2(l=reg_param),
                input_shape=(n_features,)))
model.add(Dense(512, activation="relu",
kernel_initializer="he_normal",
                kernel_regularizer=l2(l=reg_param)))
model.add(Dense(256, activation="relu",
kernel_initializer="he_normal",
                kernel_regularizer=l2(l=reg_param)))
model.add(Dense(128, activation="relu",
kernel_initializer="he_normal",
                kernel_regularizer=l2(l=reg_param)))
model.add(Dense(64, activation="relu",
kernel_initializer="he_normal",
                kernel_regularizer=l2(l=reg_param)))
```

```
model.add(Dense(32, activation="relu",
kernel_initializer="he_normal",
                kernel_regularizer=l2(l=reg_param)))
model.add(Dense(16, activation="relu",
kernel_initializer="he_normal",
                kernel_regularizer=l2(l=reg_param)))
model.add(Dense(8, activation="relu",
kernel_initializer="he_normal",
                kernel_regularizer=l2(l=reg_param)))
model.add(Dense(3, activation="softmax"))
# compile the model
model.compile(optimizer='Adadelta', loss="sparse_categorical_
crossentropy", metrics=['accuracy'])
```

The preceding implementation of neural network model has the following components.

- A *dense* input layer with 1024 nodes. Seven hidden layers with 512, 256, 128, 64, 32, 16, and 8 nodes respectively. These nodes are called *dense* because every node in these layers is connected to every node in the previous and next layer providing dense connectivity (see Figure 3-22). Finally, there's an output layer with three nodes. All nodes (except for the output layer) have the ReLU activation function. The weights on input and hidden layers are initialized using the he_normal initializer.

- Regularization occurs when a penalty is imposed for large changes in the model parameters (weights and biases) during the model training. It provides a safeguard against overfitting. The extent of the penalty imposed can be tuned using a regularization

parameter, which is usually a small numerical value. In
the MLP model shown, we used L2 regularization with
a regularization parameter of 0.02.

- The output layer in the model uses the softmax
  activation function to output estimated classification
  probabilities.

- The model uses the Adadelta optimizer, which uses a
  stochastic gradient descent method with an adaptive
  learning rate. While some other optimizers require
  a predefined learning rate, the Adadelta optimizer
  computes the learning rate adaptively. The optimizer
  uses sparse_categorical_crossentropy as the loss
  function, which is a variation of previously discussed
  categorical cross-entropy for the scenario, where each
  object can belong to only one of the sample classes.

The following code snippet is used for training the preceding MLP model.

```
# Early-stopping callback using validation
earlystop_callback = EarlyStopping(monitor='val_accuracy',
    min_delta=0.0001, patience=50)
# Save the best model
ckpt_path = './models/dnn.h5'
ckpt_callback = ModelCheckpoint(filepath=ckpt_path, mode="max",
                                monitor='val_accuracy', verbose=1,
                                save_best_only=True)
# Train the model
history = model.fit(X_train, y_train, epochs=500, batch_
size=BATCH_SIZE, validation_data=(X_valid, y_valid),
                    callbacks=[earlystop_callback,
                    ckpt_callback],
                    verbose=1)
```

Important components of the model training process are as follows.

- **Epochs:** An epoch corresponds to the number of training iterations it takes for the algorithm to process all the samples provided in the training data once. The MLP model here is expected to train for 500 epochs.

- **Early-stopping:** This concept uses a validation dataset to constantly monitor the training process. The min_delta parameter defines a minimum validation accuracy improvement. If model accuracy does not improve beyond min_delta within the number of training epochs defined by the patience parameter, training is stopped. This code implementation always saves the model whenever the previous best validation accuracy is improved. The model that provided the best validation accuracy at the end of training can be used later for generating predictions.

Figure 3-24 shows the evolution of loss function for the training and validation dataset over the training epochs. The MLP model trained with the process was used for generating predictions on the test dataset. The model generated predictions for salt, fault, and other classes with the following accuracy.

Test Accuracy: 0.713

The model accuracy can be improved to some extent by optimizing hyperparameters.

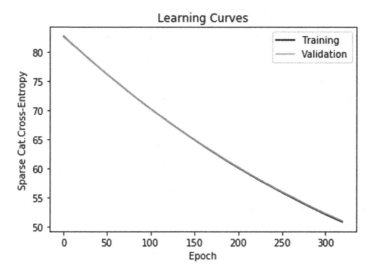

Figure 3-24. *Learning curve showing training loss for a multilayer perceptron with the number of epochs*

This example shows us that MLP can generate a model for identifying images, which was correct about 70% of the time. However, MLP is not an algorithm specialized for the image recognition task. Next, we discuss a deep neural network architecture, which is specialized in image recognition and other generic computer vision applications.

Convolutional Neural Network (CNN)

Visual identification of objects is only possible when image pixels are seen together in a 2D or 3D context. We used MLP for image classification tasks. Although MLP performed better than the flip of a coin, it did not generate predictions, which can be used in any realistic application. This lower accuracy stems from the fact that MLP looks at the images without a spatial context.

Convolutional neural networks, or CNNs, are specialized neural networks that learn to identify objects by performing computations, which allow inclusion of spatial context in the learning process. CNNs are built using convolutional layers. While most of the aspects of training a CNN remain similar to MLP training, the neural network architecture differs significantly.

Figure 3-25 summarizes some important concepts related to CNN architecture. In the convolutional layers of a CNN, convolutional filters traverse the images to extract the image features, which helps identify the class that a given image belongs to.

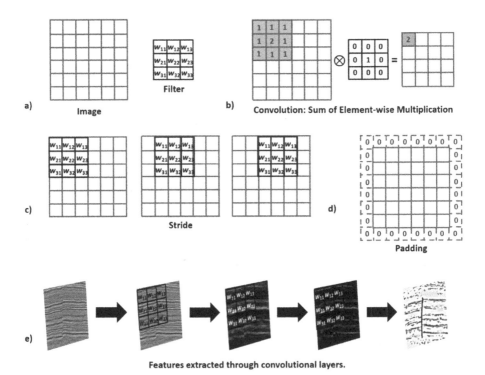

Features extracted through convolutional layers.

Figure 3-25. *Schematic depiction of (a) an image and convolutional filter, (b) convolutional operation, (c) stride, (d) zero padding around the image, and (e) conceptual representation of feature extraction happening over convolutional layers*

The following are some of the important concepts related to the CNN, as shown in Figure 3-25.

- **Filter:** A square weight matrix often represents a filter in a convolutional layer. One of the main objectives of the CNN training process is to learn optimal weights for the convolutional filters.

- **Convolution:** During the convolution operation, a sum of element-wise multiplication between the overlapping pixels of the image and the filter is computed. This value is assigned to the central pixel of the overlapping image patch (see Figure 3-25-b).

- **Stride:** The filter can traverse the image by moving one pixel at a time, or it may skip a few pixels in between. The number of pixels that filter moves at a time while traversing over the image is called *stride* (see Figure 3-25-c).

- **Padding:** When a filter traverses over the image, it reduces the size of the image because of the non-availability of overlapping cells at the edges. In certain scenarios, where it is desirable to preserve the image size, the image can be padded on the edges (see Figure 3-25-d) with zeros. For an image of size $W \times W$, being traversed by a filter of size $F \times F$, with a traversal stride S, and P cells padding the image edges, the size of the output image can be calculated by

$$\left\lceil \frac{W - K + 2P}{S} \right\rceil + 1.$$

As an image is processed through multiple convolutional layers, the filters bring out distinguishing features in the image that make it easier to identify one class from the other classes present in the sample data (see Figure 3-25-e). The following code snippet shows the CNN model that is used for classifying images in the SFO dataset.

```
model = Sequential()
model.add(Conv2D(96, (5, 5), activation="relu",
                 input_shape=(IMG_DIM, IMG_DIM, 1)))
model.add(MaxPooling2D((2, 2)))
model.add(Conv2D(128, (5, 5), activation="relu"))
model.add(MaxPooling2D((2, 2)))
model.add(Conv2D(128, (5, 5), activation="relu"))
model.add(MaxPooling2D((2, 2)))
model.add(Conv2D(64, (5, 5), activation="relu"))
model.add(Flatten())
model.add(Dense(64, activation="relu"))
model.add(Dense(32, activation="relu"))
model.add(Dense(3, activation="softmax"))
```

The CNN model in the code has alternating 2D convolutional layers and max-pooling layer. During the max-pooling operation, the algorithm selects maximum value from a set of pixels to replace the existing pixel value. The presented CNN architecture has four 2D convolution layers with a 5×5 filter size and ReLU activation function. The convolutional layers are followed by two dense layers with ReLU activation function, and an output layer with three nodes employing softmax activation. The CNN was trained using the same mechanism, which was described earlier for training the MLP. Figure 3-26 shows evolution of loss function over the training epochs. The CNN model shows high predictions accuracy in identifying salt, fault, and other classes.

```
Test Accuracy: 0.953
```

With this, we end the overview of deep learning algorithms for image classification tasks. Next, we discuss deep learning algorithms for time series forecasting.

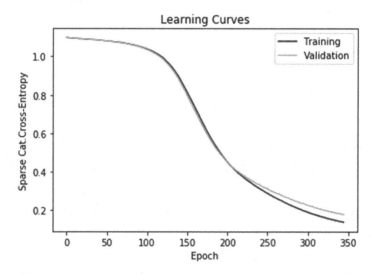

Figure 3-26. *Learning curve showing training loss for Convolutional Neural Network with the number of epochs*

Recurrent Neural Network (RNN)

Recurrent neural networks, or RNNs, provide specialized architecture, which can model sequences, including time-series data. Figure 3-27 shows the schematic representation of RNN architecture.

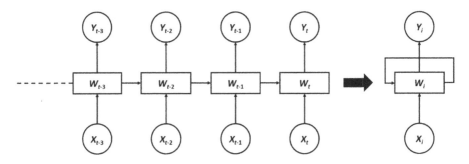

Figure 3-27. *Schematic representation of an RNN architecture*

RNN is a sequence of basic neural network models, where each model learns from the input and output of an individual step in the sequence. If we fold the models in this sequence into a single model, which takes its output as input, and keeps on updating the model parameters, we get the RNN architecture as shown in Figure 3-27 (extreme right).

However, this approach to learning model parameters (weights and bias) suffers from stability problems. Sometimes, the changes in model weights from one step in the sequence to the next step in the sequence may be too large due to large error gradients. Accumulation of gradients over multiple steps in the sequence might lead to very large updates in weights, resulting in values beyond the numeric limit that the computer can handle. This is called an *exploding gradient* problem.

On the other hand, if the error gradients are too small over multiple steps in the sequence, accumulating these gradients might lead to a value close to zero. In this scenario, the model stops updating the weights and is not able to learn. This is called a *vanishing gradient* problem.

There are specialized architectures that address exploding and vanishing gradient problems. Before getting into the discussion of one of those specialized algorithms, we look at the time series data that we use for building a time series forecasting model.

Figure 3-28 shows daily production (left) and cumulative production (right) data from two wells of the Volve dataset made available by Equinor (`https://data.equinor.com`). We are going to use this data for generating day-ahead production estimates using the production data from recent weeks.

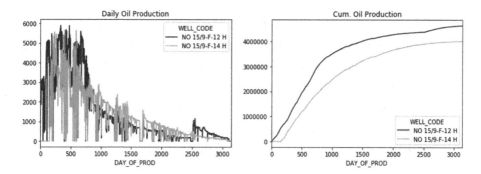

Figure 3-28. *Plots showing daily oil production (left), and cumulative oil production (right) data from two wells selected from the Volve dataset*

Long Short-Term Memory (LSTM)

We discussed the exploding and vanishing gradient problems encountered during the training of RNN models. Long short-term memory (LSTM) is an architecture, which addresses these problems by controlling the flow of information from one step in the sequence to the next. LSTM model is represented as a cell (see Figure 3-29) with information flow regulating *gates*. The cell state variable in the LSTM model retains long-term memory of the sequence, or the time series being modeled. These gates include the following.

- **Forget gate:** This gate gathers the information learned based on the combination of input to the cell, and output (hidden state) from the previous step in the sequence. Then, using a sigmoid function, it determines the amount of information that should

143

be discarded. The sigmoid function provides values
between 0 and 1, with 1 depicting retention of all
the information, and 0 depicting discarding all the
information.

- **Input gate:** This gate also uses input to the cell, and
output from the previous step. The input gate uses a
sigmoid function to learn the amount of information
that should be retained for updating the cell state. The
same input is passed through *tanh* activation. The
input gate provides values between 0 and 1, whereas
tanh provides values between –1 and 1. A combination
of these two is used for updating the cell state.

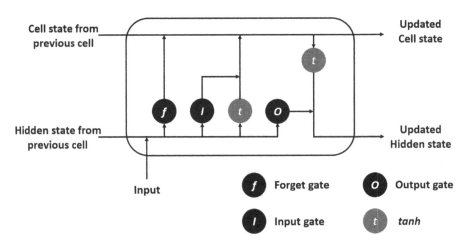

Figure 3-29. *Schematic depiction of long short-term memory (LSTM)
cell architecture*

- **Output gate:** The output gate uses an input similar
to the previous two gates. Using a sigmoid function,
it learns the amount of information necessary to
compute the model output. A combination of *tanh*

activation applied to the cell state (long-term memory
of the sequence) and values received from the output
gate provides cell output.

The following code snippet shows a sample implementation of LSTM
architecture using TensorFlow. Cumulative production data from the past
14 days is used for predicting cumulative production on the next day. This
implementation contains two stacked LSTM layers, followed by a dense
layer, and an output layer with a single node (one output variable). We
use the Adadelta optimizer and minimize the mean-squared error (MSE)
metric for training the LSTM model. The implementation also uses early-
stopping using validation data (not shown here).

```
model = Sequential()
model.add(LSTM(NUM_HIDDEN, activation="relu", return_
sequences=True, input_shape=(LOOKBACK, 1)))
model.add(LSTM(NUM_HIDDEN, activation="relu", return_
sequences=False))
model.add(Dense(NUM_HIDDEN))
model.add(Dense(1))
model.compile(optimizer='Adadelta', loss="mse")
```

Figure 3-30 (left) shows the normalized cumulative production for
the observed test data, and the cumulative production predicted by the
LSTM model. Figure 3-30 (right) shows a cross plot of observed test data,
and LSTM predicted values for the test data. The results indicate good
prediction capability of the LSTM model, after some mismatch during
the early days of production from the well. An industrially relevant
implementation outputs production for several days—something we
encourage you to implement using the provided code.

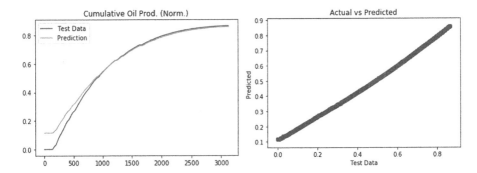

Figure 3-30. *Plots showing test data and corresponding LSTM model predictions for cumulative oil production(left), and a cross plot of observed and LSTM predicted cumulative oil production for the test data (right)*

Hyperparameter Optimization

We have mentioned that hyperparameter optimization can help to improve model accuracy. In this section, we demonstrate a simple code example of the optimization process. But, before getting into the implementation details, let's learn the definition of hyperparameters.

What Are Hyperparameters?

Hyperparameters are the variables, which are used during the machine learning algorithm-training process to learn the best model parameters. In other words, hyperparameters are the parameters, which help in learning optimal model parameters. Some examples of hyperparameters include the following.

- The number of layers in a neural network

- The number of nodes in the neural network layers

- The type of activation function for the neural network nodes

- The number of iterations or epochs for model training

- The choice of "kernel" type in kernel-based methods

- The regularization parameter

- The type of neural network weight initialization

- The tolerance for early-stopping (`min_delta`)

- The number of epochs to wait before early-stopping (`patience`)

While the name of this approach suggests that it is only applicable for optimizing hyperparameters, in certain scenarios, it can also tune model parameters. We demonstrate this using a code example.

Using Hyperopt for Hyperparameter Optimization

We are using Hyperopt, a Python library, to demonstrate the hyperparameter optimization process. We demonstrate the process using a support vector classifier algorithm. We used the lithotype classification data and built a vanilla classifier. Accuracy of this vanilla classifier serves as a baseline to evaluate the improvement in model performance as a result of the hyperparameter optimization process. The following code snippet shows vanilla classifier implementation and corresponding classification accuracy.

```
svc = SVC()
svc.fit(X_train, y_train)
# Prediction on test data
y_pred = svc.predict(X_test)
# Accuracy Metrics
clf_metrics(y_test, y_pred)
```

Classification Accuracy Score: **0.8621700879765396**

Hyperopt uses an objective function, which is minimized to compute optimal values of model parameters and hyperparameters. In the following example, an accuracy score from k-fold cross-validation multiplied by –1, is used as the objective function. The objective function is minimized by iteratively adjusting the hyperparameters and model parameters over 50 iterations. Some important aspects of the implementation include the following.

- **K-fold cross-validation:** In this method, training data is split into k subsets or folds. By using these k-folds, k separate models are trained. Each model is trained by holding one of the k-folds for validation (validation fold) at a time and using remaining k-1 folds for training the model. The accuracy score for k-fold cross-validation is computed by averaging accuracy scores of individual models on the data in the respective validation fold.

- **Parameter space:** Defines the range of values that model parameters and hyperparameters associated with the model can assume.

- **Hyperparameters:** In this example, the type of kernel is a hyperparameter. The constant C, degree of the polynomial, and gamma are model parameters. This is an example where both hyperparameters and model parameters are being optimized at the same time.

The following code snippet shows the approach for finding an optimal kernel, and corresponding parameters for the support vector classifier model.

```
def objective(params):
    svc = SVC(**params)
    return -1. * cross_val_score(svc, X_train, y_train).mean()
```

```
kernels = ['rbf','poly','rbf','sigmoid']
space = {'C':hp.lognormal('C', 0, 1),
         'kernel':hp.choice('kernel', kernels),
         'degree':hp.choice('degree', range(1, 15)),
         'gamma':hp.uniform('gamma', 1e-2, 1e2)
         }

trials = Trials()
best_svc = fmin(objective, space, algo=tpe.suggest, max_
evals=50, trials=trials)
print(best_svc)
```

**{'C': 1751.5444736790396, 'degree': 11, 'gamma':
5.899864955354083, 'kernel': 0}**

After the determination of the best kernel type, and model parameters, the best model is used for generating predictions on the test data.

```
svc = SVC(C=best_svc['C'],
          kernel=kernels[best_svc['kernel']],
          degree=best_svc['degree'],
          gamma=best_svc['gamma'])
svc.fit(X_train, y_train)
# Prediction on test data
y_pred = svc.predict(X_test)
# Accuracy Metrics
clf_metrics(y_test, y_pred)
```

Classification Accuracy Score: **0.9618768328445748**

The example shows that hyperparameter optimization using the Hyperopt library improves model classification accuracy from 0.86 to 0.96. A similar process can be applied for improving the accuracy of any other machine learning or deep learning algorithm by implementing an appropriate objective function.

Summary

I didn't come here to tell you how this is going to end. I came here to tell you how it's going to begin.

—Neo in *The Matrix* [12]

Machine learning and deep learning are vast topics. In this chapter, we attempted to provide an overview of several concepts related to machine learning and deep learning algorithms. Our focus was on providing simple code implementation of each algorithm using open source datasets. However, we acknowledge that mastering these concepts requires further study and continued practice. Please refer to the next steps recommended in the "Further Reading" section. In the following chapters, we are going to build some interesting machine learning applications for oil and gas industry-related problems by applying the concepts that we discussed in this chapter.

Acknowledgments

We thank Equinor AS, the former Volve license partners ExxonMobil Exploration and Production Norway AS and Bayerngas (now Spirit Energy), for permission to use the Volve dataset and the many people who have contributed to the work there. Please visit data.equinor.com for more information about the Volve dataset and license terms of use.

We also thank TGS for permission to use the dataset from the TGS Salt Identification Challenge on Kaggle (`www.kaggle.com/c/tgs-salt-identification-challenge/`). Also, we acknowledge New Zealand Petroleum and Minerals (`www.nzpam.govt.nz/cms`) for providing the data.

Finally, we would like to thank Bhaskar Mandapaka for some very interesting and insightful discussions, which helped shape this chapter.

FURTHER READING

Now that we have finished the overview of machine learning and deep learning, we recommend further reading.

- *Deep Learning* (www.deeplearningbook.org) [13] is a free online book that provides an in-depth discussion of theoretical and mathematical foundations of machine learning and deep learning algorithms.

- Please thoroughly read the online documentation of the machine learning libraries used in this chapter. It will help you get a better grasp of different parameters related to each algorithm in the corresponding library.

References

[1] T. Mitchell, *Machine Learning*, McGraw Hill, 1997.

[2] P. Domingos, "A Few Useful Things to Know About Machine Learning," *Communications of the ACM,* vol. 55, no. 10, pp. 78–87, October 2012.

[3] R. Mooney, "Machine Learning Introduction," the University of Texas at Austin, [Online]. Available: https://www.cs.utexas.edu/~mooney/cs391L/slides/intro.pdf.

[4] J. Brownlee, "How to Reduce Generalization Error With Activity Regularization in Keras," November 2018. [Online]. Available: https://machinelearningmastery.com/how-to-reduce-generalization-error-in-deep-neural-networks-with-activity-regularization-in-keras/.

[5] S. Raschka, "About feature scaling and normalization and the effect of standardization for machine learning algorithms," *Polar Political Legal Anthropology Rev.,* vol. 30, no. 1, pp. 67–89, 2014.

[6] J. Brownlee, "What is the Difference Between Test and Validation Datasets?," July 2017. [Online]. Available: `https:// machinelearningmastery.com/difference-test-validation-datasets/.`

[7] G. James, D. Witten, T. Hastie, and R. Tibshirani, *An Introduction to Statistical Learning with Applications in R*, Springer, 2013.

[8] H. J. Seltmann, "Experimental design and analysis," [Online]. Available: `http://www.stat.cmu.edu/~hseltman/309/Book/Book.pdf.`

[9] A. Andrade, "Exploratory Data Analysis," [Online]. Available: `https://datascienceguide.github.io/exploratory-data-analysis.`

[10] T. Hastie, R. Tibshirani, and J. Friedman, *The Elements of Statistical Learning*, New York: Springer, 2001.

[11] V. Sze, Y. H. Chen, T. J. Yang, and J. Emer, "Efficient Processing of Deep Neural Networks: A Tutorial and Survey," *Proceedings of the IEEE,* vol. 105, no. 12, 2017.

[12] *The Matrix,* 1999.

[13] I. Goodfellow, Y. Bengio, and A. Courville, *Deep Learning*, MIT Press, 2016.

## CHAPTER 4

Geophysics and Seismic Data Processing

In the previous chapters, we gained an understanding of Python programming, learned concepts of machine learning and deep learning, and implemented some of these concepts using Python. The focus of this chapter is to give an overview of the applications of machine learning in the field of geophysics. The primary objective of exploration workflows is the use of seismic data processing to build earth models to estimate the reservoir properties. This is a mature and well-studied problem, with several decades of academic and industrial research behind the science of seismic imaging and reservoir property estimation.

Machine Learning in exploration geophysics has a long history almost coinciding with the rise and ebb in machine learning in the last few decades. Applications in exploration geophysics workflows are many and plenty. It is beyond the scope of this chapter to explain each of them in detail. The approach of this chapter is to offer an overview of the seismic imaging process, and discuss in detail about the challenging problem of salt interpretation. The chapter will conclude with a brief summary of several other problems and pointers to relevant literature in exploration and interpretation workflows.

© Yogendra Narayan Pandey, Ayush Rastogi, Sribharath Kainkaryam,
Srimoyee Bhattacharya, and Luigi Saputelli 2020
Y. N. Pandey et al., *Machine Learning in the Oil and Gas Industry*,
https://doi.org/10.1007/978-1-4842-6094-4_4

Seismic Reflection Experiment

Reflections and refractions in seismic waves within the earth were first observed on seismographs that recorded earthquake-generated seismic waves. The general principle of a seismic reflection experiment is to send waves into the subsurface and record them with several thousands of sensors. Figure 4-1 shows a schematic of an experiment that is used to acquire reflection data in the marine setting. An explosive source of energy is used to emit waves that are reflected from contrasts in the subsurface. These reflections from key layers are recorded by sensors. Several thousands of sensors are towed by the streamer as shown in the cartoon.

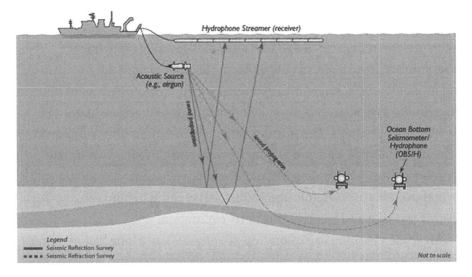

Figure 4-1. *Schematic of a marine seismic reflection experiment (Courtesy: National Science Foundation)*

Figure 4-2 shows an example of a sample record that is recorded by a seismic experiment. A data record of this type is called a "shot record" as it corresponds to a single explosion from which the reflection waves have been recorded. For a typical marine experiment, there are several thousands of these shots acquired over an area.

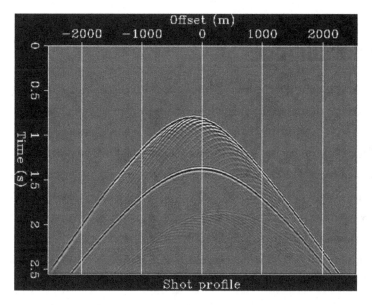

Figure 4-2. *Shot gather: Data captured by a single seismic experiment (Courtesy: Introduction to Seismic Imaging)*

The objective of seismic processing is to use sensor data, as shown in Figure 4-2, to generate an image of the subsurface, as well as estimate the properties of the Earth. Typically, the objective is to estimate the velocity of the compressional waves traveling through the Earth as these are a proxy for the reservoir properties at the scale of the wavelength. The process of converting these shot records to generate an image of the earth is called imaging. During the process of generating a subsurface image, it is also essential to estimate the properties of the subsurface through a process called earth model building. An overview of the seismic acquisition problem and its various challenges is explained in reference [1].

Inverse Problem: Imaging and Velocity Model Building

The goal of recording petabytes of seismic data is to obtain an image of the subsurface of the earth, as well as estimate the physical properties of the subsurface. A minimal model for modeling wave propagation in the subsurface is given by the acoustic wave equation. More complicated models exist for accurately modeling wave propagation. Such models are beyond the scope of this introductory chapter. An excellent mathematical overview of the wave equation based imaging, and the inverse problem associated with it can be found in reference [2].

A model for acoustic wave equation can be defined by the partial differential equation system shown in Figure 4-3.

$$\nabla^2 u(\mathbf{x}, t) - \frac{1}{c^2} \frac{\partial^2 u(\mathbf{x}, t)}{\partial t^2} = f(\mathbf{x_s}, t); \text{ i.c.; b.c.}$$
$$\mathcal{M} = \text{a set of models}$$
$$\mathcal{D} = \text{a space of data}$$
$$\text{Forward Operator } \mathcal{F} : \mathcal{M} \rightarrow \mathcal{D}$$

Figure 4-3. *Acoustic Wave Equation*

In the above system of equations, $u(\mathbf{x}, t)$ is the pressure field that is recorded by the hydrophones, $c(\mathbf{x})$ is the wavespeed of sound waves (compressional waves) in the subsurface and $f(\mathbf{x}_s, t)$ is the source signature for the experiment described in the previous section. The system is solved with an appropriate set of initial conditions and boundary conditions.

In the mathematical and geophysical literature, this linear system is called a forward problem as it gives the definition of a physical system for modeling the wave field, given the definition of wave speed in the earth's subsurface. This is a well-studied and well defined problem with several centuries of academic history behind it. The interested reader is suggested to consult reference [3], and the references therein.

However, the objective of seismic imaging is to solve the inverse of the problem as defined by the system above. The objective is to find a model that explains the data that has been recorded (Figure 4-4). Unlike the forward problem which is a linear problem, the inverse problem is a strongly nonlinear problem that is challenging to solve. Further, the solution to the inverse problem is non-unique. Non-uniqueness implies several models can reasonably explain the data that has been recorded.

$$\left(\frac{1}{c_0^2(\mathbf{x})}\frac{\partial^2 u}{\partial t^2}(\mathbf{x}, t) - \nabla^2\right)\delta u(\mathbf{x}, t) = \frac{2\delta c(\mathbf{x})}{c_0^3(\mathbf{x})}\frac{\partial^2 u}{\partial t^2}(\mathbf{x}, t)$$
$$F[c_0]\delta c(\mathbf{x}) = \delta u(\mathbf{x})$$
$$Am = d$$

Figure 4-4. *Inverse problem*

Figure 4-5 shows the comparison of two broad classes of problems that form the discipline of seismic imaging. The problem of imaging is accurate positioning of reflectors while the problem of inversion is estimating a subsurface model (velocity and anisotropic properties) that explain the data that has been recorded in the subsurface.

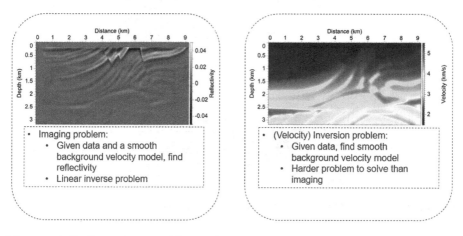

Figure 4-5. *Imaging and Inversion*

Physical Properties of Sedimentary Rocks

In sedimentary rocks, compressional velocity varies significantly, and heterogeneity is observed on all scales. Figure 4-6 shows the velocity profile from a well from onshore United States. The variation in velocity can be observed on various scales, varying across several orders of magnitude - from micrometers to millimeters to kilometers.

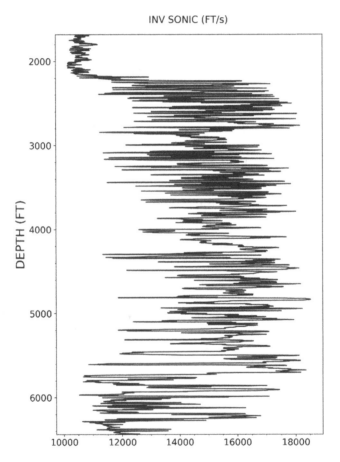

Figure 4-6. *Multiscale heterogeneity of rock properties (Data courtesy: TGS)*

In order to study the nonlinear inverse problem defined in the equation system shown in Figure 4-4, a perturbation approach is used to study the scattering caused by the subsurface. Born approximation is a technique originally used in quantum mechanics to study scattering by an extended body. It is shown to be accurate when the size of the scatterers is small compared to the wavelength. Interested readers are referred to the reference [2].

Velocity model can be separated according to scales. A smooth macromodel that contains the long-scale component of velocity. Typically, this is in the form of a smooth background. The scatterers are represented in the form of an oscillatory perturbation above the smooth background, which represents the high-frequency component of the velocity. This model of the subsurface models single scattering only. In other words, these model the primary reflected waves. A careful treatment of Born approximation is beyond the scope of this introductory discussion on seismic imaging. The interested reader is referred to reference [2] for details.

Imaging and Inversion

The objective of this section is to offer an insight into the process of velocity estimation, and imaging scatterers in the subsurface. Given the data recorded on the subsurface and a smooth background velocity model, the goal of imaging is to find the reflectivity of the subsurface. This is a linear inverse problem, and several methods exist to image the scatterers in the subsurface.

To accurately identify the reflectors in the subsurface, it is also essential to have an accurate velocity model. Given the data recorded on the surface, the goal of velocity inversion is to find a smooth background velocity model that explains the data recorded on the surface. This is a much harder problem than imaging, and is strongly nonlinear. In practice, the process of estimating reflectors and updating the velocity model is solved in an iterative loop till certain criteria are met.

The linear inversion problem for positioning the reflectors in the subsurface is approximated in practice. The mathematical details are beyond the scope of this introductory material on imaging. An approximate inverse is computed and this process is called as migration. There are several migration algorithms that are currently used, including, Kirchhoff migration, reverse time migration, wave equation migration, etc. The goal is to backward propagate the receiver wavefield and forward propagate the source wavefield, and the image is the cross-correlation of the wavefields at various values of timelag. The resulting zero lag value is the image (Figure 4-7).

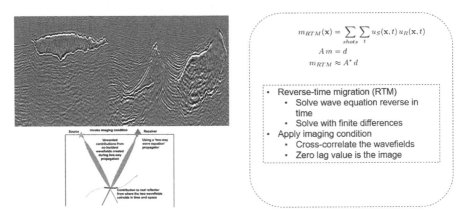

$$m_{RTM}(\mathbf{x}) = \sum_{shots} \sum_{t} u_S(\mathbf{x}, t)\, u_R(\mathbf{x}, t)$$

$$A\,m = d$$

$$m_{RTM} \approx A^* d$$

- Reverse-time migration (RTM)
  - Solve wave equation reverse in time
  - Solve with finite differences
- Apply imaging condition
  - Cross-correlate the wavefields
  - Zero lag value is the image

Figure 4-7. *Overview of migration*

A common technique to estimate the velocity properties of the subsurface is to use a method called as Full Waveform Inversion. It is an iterative method to match the data recorded on the surface in a certain frequency range. This technique computes the smooth part of the velocity model, and like most strongly nonlinear optimization problems is very sensitive to the choice of the initial model. Owing to the bandlimited nature of the data, this approach results in a smooth solution and cannot resolve structures on all scales. Reference [4] describes the challenges of building a velocity model in the Gulf of Mexico. A velocity model obtained from Full Waveform Inversion at Atlantis Field in Gulf of Mexico from reference [4] is shown in Figure 4-8.

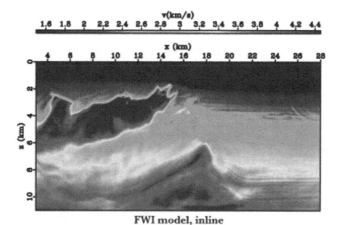

Figure 4-8. *Velocity model derived from full waveform inversion [4]*

Due to the bandlimited nature of the data, an accurate starting model is important to solve this optimization problem to avoid being stuck at a local minima. The optimization problem is formulated with the assumption that the observed and predicted waveforms are within half a wavelength at the lowest frequency. However, lack of low frequency information in seismic data results in cycle-skipping, which results in the problem getting stuck at local minima instead of reaching the global minima. This problem is harder to solve in basins, such as, the Gulf of Mexico, which contains velocity contrasts owing to the presence of complex structures of salt that make the optimization problem challenging to solve (Figure 4-9).

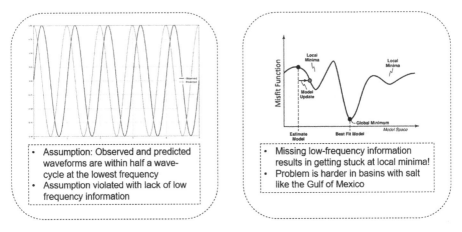

Figure 4-9. *Challenges of lack of low frequency*

Overcoming Low Frequency Challenges

Over the last several years, there has been rapid technological growth in seismic acquisition technology. There has been a marked improvement in the range of coverage and azimuths (angles) over the course of last two decades -- near azimuth acquisition, wide azimuth acquisition, full azimuth acquisition are some examples of the operational innovations over the last couple of decades. Further, there have been several attempts to broaden the frequency range of acquisition over the last several years. Low frequency data is crucial as it helps avoid the optimization problem being stuck at a local minima. Hence, several attempts have been made in the last few years to acquire low frequency data. But, acquisition technologies are extremely expensive and can be a challenge to implement on a sufficiently large scale. However, in cases where low frequency data is not available, starting with better initial velocities and having an *a priori* structure of the velocity model helps in accelerating the convergence.

Salt interpretation is a challenging task in the building of velocity models for building the structure. The state-of-the-art methods in salt interpretation involve using manual interpretation by a trained team of geologists and domain experts, who serve to impose their knowledge of geological priors on the inverse problem. However, given the iterative nature of the inverse problem, it is required for the interpreters to pick the structure of the salt after every single application of imaging. The process of picking the structure of the salt is a time intensive problem that is critical to building accurate seismic velocity models.

Salt Interpretation with Machine Learning

In this section, the focus will be on giving an overview of supervised learning problems in seismic imaging. As an example, we will focus in detail on the problem of binary segmentation for the case of velocity model building described in the previous sections.

The problem of semantic segmentation is well studied in the computer vision community. The goal is to assign each pixel of the image an object class. Figure 4-10 shows examples from a computer vision dataset and seismic image-mask pair to draw similarity between the problems. The left pane shows two samples from a semantic segmentation dataset. The right pane shows the seismic image and the mask obtained by semantic segmentation process on the given input image. Apart from recognizing various objects in the image, the task is also to delineate the boundaries of the object.

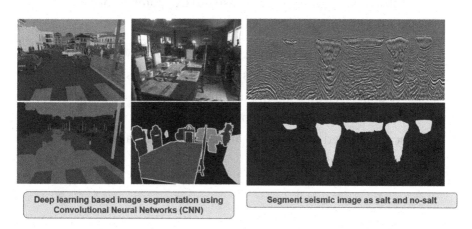

Deep learning based image segmentation using Convolutional Neural Networks (CNN)

Segment seismic image as salt and no-salt

Figure 4-10. *Semantic segmentation: Computer vision v/s Seismic Imaging*

Conceptually, this is similar to the interpretation process used in velocity model building workflow. Given a seismic image, the goal is to classify each pixel in the seismic image as salt or sediment. This can be cast as a supervised learning problem similar to the semantic segmentation problem for computer vision, where the image is the output from the migration algorithm, and the label is a binary mask.

An example is shown in Figure 4-10 that shows the similarity of the semantic segmentation problem in seismic imaging. This is a slice of a larger 3D cube of data that represents the post-migrated amplitudes. The labels are extracted from the velocity model of the same area. An important point to note is that these are continuous amplitudes unlike images in computer vision datasets that have been quantized. Further, images sampled from different migrated volumes have different amplitude distributions. Therefore, it is common to apply normalization of some form before applying deep learning algorithms on these images.

There are some important aspects of applying deep learning algorithms to seismic images. Please note that these are specific to the problem of interpreting salt from seismic images.

1. **Choice of image sizes:** A suitable window extraction algorithm can be used to use patches for training or inference.

2. **Preprocessing of the images:** In order to convert the continuous amplitudes to quantized levels, we apply normalization. Most methods of normalization do work in principle. However, scaling based on the mean and standard deviation has shown to work.

Dataset Description

We use the dataset from TGS Kaggle Salt Interpretation challenge [5]
that was hosted by TGS and Kaggle. The data is a set of images chosen at
various locations chosen at random in the subsurface. The images are 101
× 101 pixels, and each pixel is classified as salt or sediment. Figure 4-11
shows a few samples of the images from the chosen dataset.

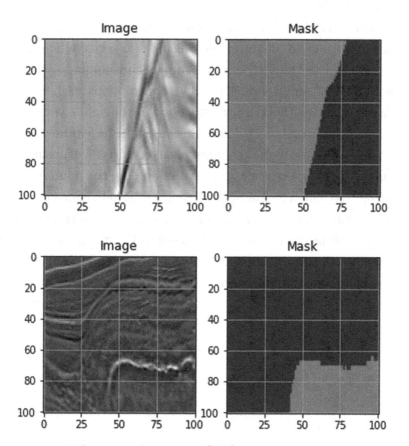

Figure 4-11. *A few sample images (left) and corresponding labels
(right) from the Kaggle dataset*

U-Net for Semantic Segmentation

In order to train a deep learning model to perform semantic segmentation, several architectures have been published over the last few years. Some of the architectures for semantic segmentation have established state-of-the-art records in segmentation [6]. Although we will list the publications explaining the models in detail, in this chapter, we will focus on U-Net as a canonical example [7].

As shown in Figure 4-12, U-Net consists of a contracting path (as seen on the left of the image), and an expanding path (as seen on the right). The contracting path follows the architecture of convolutional neural networks. Repeated application of two 3 × 3 convolutions, each followed by a Rectified Linear Unit (ReLU), and a max-pooling operation with a stride of 2 for downsampling. At each downsampling step, the feature channels are doubled. Every step in the expanding path consists of an upsampling of the feature map, followed by a 2 × 2 convolution ("up-convolution"). This upsampling step halves the number of feature channels, concatenates it with the corresponding cropped feature map from the contracting path, two 3 × 3 convolutions each followed by a ReLU. The cropping is necessary due to the loss of border pixels in every convolution. At the final layer a 1 × 1 convolution is used to map each 64-component feature vector to the desired number of classes. In total the network has 14 convolutional layers.

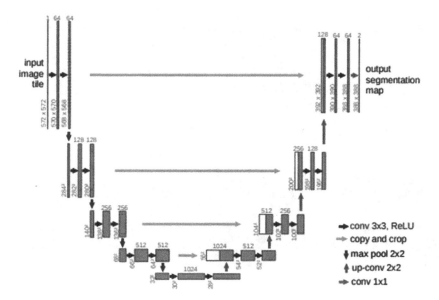

Figure 4-12. *U-Net architecture (Reference [7])*

Loss Function for U-Net

An objective function of intersection over union (IOU) is commonly used for evaluating the accuracy of semantic segmentation problems (Figure 4-13). Several other objective functions have been proposed in the computer vision community. We want to note that although IOU is a commonly used metric for evaluating segmentation networks, the example shared in the Jupyter notebook uses binary crossentropy as the loss function for the simple reason that it is easier to understand and commonly available in most of the deep learning libraries.

To apply the metric, the intersection over the predicted object and the expected object are computed for the numerator, and the union over the predicted object and the expected object are computed for the denominator. The quotient is the metric for measuring the accuracy. As can be seen, if the model predicts the shape of the object accurately, the intersection over union will be 1.

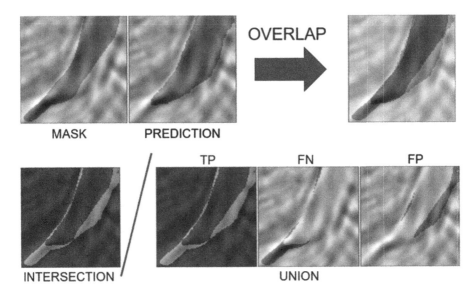

Figure 4-13. *Objective function of intersection over union*

U-Net Implementation

A code snippet representing a U-Net model is shown here. It consists of a contracting path and an expanding path, similar to the architecture shown in Figure 4-12. The main idea behind U-Net is to supplement the contracting layers, where the contraction that happens by pooling are replaced either by upsampling or transposed convolution layers. This is defined in the model architecture listed in the following code:

```
def UNet(input_image, n_filters=16, dropout=0.5,
batchnorm=True):
    # Encoder
    c1 = conv2d_block(input_image, n_filters=n_filters*1,
    kernel_size=3, batchnorm=batchnorm)
    p1 = MaxPooling2D((2, 2)) (c1)
    p1 = Dropout(dropout*0.5)(p1)
```

```
c2 = conv2d_block(p1, n_filters=n_filters*2, kernel_size=3,
    batchnorm=batchnorm)
p2 = MaxPooling2D((2, 2)) (c2)
p2 = Dropout(dropout)(p2)

c3 = conv2d_block(p2, n_filters=n_filters*4, kernel_size=3,
    batchnorm=batchnorm)
p3 = MaxPooling2D((2, 2)) (c3)
p3 = Dropout(dropout)(p3)

c4 = conv2d_block(p3, n_filters=n_filters*8, kernel_size=3,
    batchnorm=batchnorm)
p4 = MaxPooling2D(pool_size=(2, 2)) (c4)
p4 = Dropout(dropout)(p4)

c5 = conv2d_block(p4, n_filters=n_filters*16, kernel_
    size=3, batchnorm=batchnorm)

# Decoder
u6 = Conv2DTranspose(n_filters*8, (3, 3), strides=(2, 2),
    padding="same") (c5)
u6 = concatenate([u6, c4])
u6 = Dropout(dropout)(u6)
c6 = conv2d_block(u6, n_filters=n_filters*8, kernel_size=3,
    batchnorm=batchnorm)

u7 = Conv2DTranspose(n_filters*4, (3, 3), strides=(2, 2),
    padding="same") (c6)
u7 = concatenate([u7, c3])
u7 = Dropout(dropout)(u7)
c7 = conv2d_block(u7, n_filters=n_filters*4, kernel_size=3,
    batchnorm=batchnorm)
```

```
    u8 = Conv2DTranspose(n_filters*2, (3, 3), strides=(2, 2),
        padding="same") (c7)
    u8 = concatenate([u8, c2])
    u8 = Dropout(dropout)(u8)
    c8 = conv2d_block(u8, n_filters=n_filters*2, kernel_size=3,
        batchnorm=batchnorm)

    u9 = Conv2DTranspose(n_filters*1, (3, 3), strides=(2, 2),
        padding="same") (c8)
    u9 = concatenate([u9, c1], axis=3)
    u9 = Dropout(dropout)(u9)
    c9 = conv2d_block(u9, n_filters=n_filters*1, kernel_size=3,
        batchnorm=batchnorm)

    outputs = Conv2D(1, (1, 1), activation="sigmoid") (c9)
    model = Model(inputs=[input_img], outputs=[outputs])
    return model

model.compile(optimizer=Adam(), loss="binary_crossentropy",
metrics=["accuracy"])

callbacks = [
    EarlyStopping(patience=10, verbose=1),
    ModelCheckpoint('./model/model-tgs-salt.h5', verbose=1,
    save_best_only=True, save_weights_only=True)
]

results = model.fit(X_train, y_train, batch_size=16, epochs=25,
callbacks=callbacks, validation_data=(X_valid, y_valid))
```

The Jupyter notebook accompanying this chapter provides step-by-step code for training a simple U-Net model using the dataset from Kaggle.

Interpretation Results

We conclude this practical example with a brief discussion of results. The network was trained with Adam optimizer on a dataset of 4,000 image patches, of which 10% was held out as a validation dataset. The model was trained for 25 epochs with early-stopping, and the best model was saved for generating prediction for the validation dataset. A few visualizations from the predictions generated for the samples in the validation set are shown in Figure 4-14. As can be seen from the probabilities in the two example samples, the model has learnt to predict the boundary between salt and sediment. These probabilities can be thresholded to obtain predictions for a given image patch.

An important thing to note is that the neural network model has been trained on patches. However, this trained model is often applied on migrated images. As described in the previous section, the migrated images are three dimensional volumes. In order to obtain a salt mask, inference is done patchwise and the probabilities are summed by ensuring weights are divided in an appropriate manner. This can be achieved by a simple moving window algorithm that infers each patch in a line individually. Salt mask resulting from the application of such a workflow is shown in Figure 4-15.

Generalization of deep neural networks for the purpose of seismic interpretation is a challenge. A discussion of these challenges is beyond the scope of this introductory chapter.

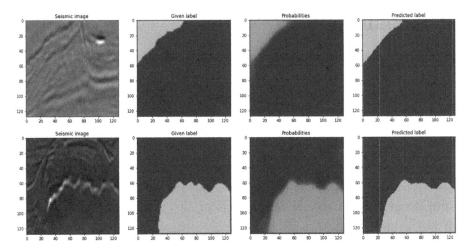

Figure 4-14. *Predictions of U-Net on the image patches. (left to right) Seismic image, recorded labels, predicted probabilities and predicted labels. Probabilities are obtained by the application of sigmoid layer and labels are obtained by thresholding the probabilities*

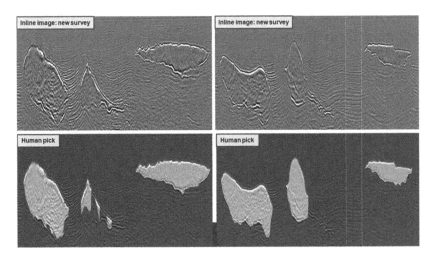

Figure 4-15. *Comparison of interpretation from U-Net and ground truth labeled by interpreters*

A Sampling of Problems

We focused on the problem of interpreting salt from seismic images in this chapter. However, several other applications of machine learning have been published in the geophysics community. We give a sampling of problems in geophysics and seismic data processing with a selection of references.

Facies classification is an important topic at the core of reservoir property estimation. It is crucial to classify lithologies and facies to build a detailed understanding of the depositional environments encountered in the wellbore. This is an extremely important problem in exploration. Several authors have explored this problem in detail. A good introduction to the application of machine learning in facies classification can be found in references [8] and [9].

Inversion for reservoir properties is another important area of application. Reference [10] provides an example using a physics-guided convolutional neural network to estimate reservoir properties. Another commonly studied problem is the process of interpreting faults from seismic images. Reference [11] presents an example to interpret faults in 3D using segmentation performed by 3D convolutional neural networks.

Summary

In this chapter, we offered an overview of geophysics and seismic data processing. We focused on the problem of salt interpretation and gave an overview of other important problems in geophysics. Given the nature of image data that is available in the seismic imaging community, the problems are amenable for application of ideas from the computer vision community. Several problems in this area are multidisciplinary, and this chapter offered an overview of a variety of them.

Acknowledgements

The author would like to thank TGS for the permission to write this chapter, and use the Kaggle data set for salt interpretation. Cen Ong, Arvind Sharma, Satyakee Sen, Cable Warren and Sathiya Namasivayam are thanked for all the useful discussions and insights.

References

[1] B. J. Evans, A Handbook for seismic data acquisition in exploration, Society of Exploration Geophysicists, 1997.

[2] W. W. Symes, Mathematics of Reflection Seismology, 1995, p. 1–85.

[3] Ö. Yilmaz, Seismic Data Analysis: Processing, Inversion, and Interpretation of Seismic Data, Society of Exploration Geophysicists, 2001.

[4] Shen, Xukai, I. Ahmed, A. Brenders, J. Dellinger, J. Etgen and S. Michell, "Salt Model Building at Atlantis With Full-Waveform Inversion," in SEG Technical Program Expanded Abstracts 2017, 2017.

[5] S. Kainkaryam, C. Ong, S. Sen and A. Sharma, "Crowdsourcing Salt Model Building: Kaggle-TGS Salt Identification Challenge," in 81st EAGE Conference and Exhibition 2019, 2019.

[6] J. Long, E. Shelhamer and T. Darrell, "Fully Convolutional Networks For Semantic Segmentation," in Proceedings of the IEEE Conference On Computer Vision And Pattern Recognition, 2015.

[7] O. Ronneberger, P. Fischer and T. Brox, "U-net: Convolutional networks for biomedical image segmentation," in International Conference on Medical Image Computing and Computer-Assisted Intervention, 2015.

[8] Y. Alaudah, P. Michalowicz, M. Alfarraj and G. AlRegib, "A Machine-Learning Benchmark for Facies Classification," Interpretation, vol. 7, no. 3, p. SE175–SE187, 2019.

[9] T. Wrona, I. Pan, R. L. Gawthorpe and H. Fossen, "Seismic Facies Analysis Using Machine Learning," Geophysics, vol. 83, no. 5, pp. O83-O95, 2018.

[10] R. Biswas, M. K. Sen, V. Das and T. Mukerji, "Prestack and Poststack Inversion Using a Physics-Guided Convolutional Neural Network," Interpretation, vol. 7, no. 3, p. SE161–SE174, 2019.

[11] X. Wu, L. Liang, Y. Shi, Fomel and Sergey, "FaultSeg3D: Using Synthetic Data Sets to Train an End-To-End Convolutional Neural Network for 3D Seismic Fault Segmentation," Geophysics, vol. 84, no. 3, pp. IM35-IM45, 2019.

[12] S. Bader, X. Wu and S. Wu, "Missing log data interpolation and semiautomatic seismic well ties using data matching techniques," Interpretation, vol. 7, no. 2, p. T347–T361, 2019.

[13] P. Jaccard, "The Distribution of the Flora of the Alpine Zone. 1," *New Phytologist,* vol. 11, no. 2, pp. 37-50, 1912.

[14] J. Vamaraju and M. K. Sen, "Unsupervised physics-based neural networks for seismic migration," *Interpretation,* vol. 7, no. 3, p. SE189–SE200, 2019.

[15] X. Wu, L. Liang, Y. Shi, Z. Geng and S. Fomel, "Deep learning for local seismic image processing: Fault detection, structure-oriented smoothing with edge-preserving, and slope estimation by using a single convolutional neural network," in *SEG Technical Program Expanded Abstracts 2019*, 2019.

CHAPTER 5

Geomodeling

Geomodeling is an important step in the process of exploration and production planning. Geomodeling is performed after developing an understanding of the structural framework of the reservoir using the information extracted during seismic interpretation. A three-dimensional (3D) grid is created by using the horizons and faults defined by the structural framework. This 3D grid is used along with the well logs to generate a detailed 3D model of the reservoir properties, such as porosity, permeability, and so forth. This step is a precursor to reservoir simulation. The calculations based on the 3D reservoir model also help in estimating important quantities, including original oil in place (OOIP).

Once a 3D grid is created using the structural framework, it requires estimates of discrete petrophysical properties, such as facies, and continuous properties, such as porosity and permeability, before performing any further workflow steps. The data from the well logs form the basis for property estimation. A wide array of modeling techniques are applied to generate these properties on every cell of the 3D grid. Some of the most common techniques involve the following algorithms [1] [2].

- Simple kriging

- Ordinary kriging

- Universal kriging

- Sequential Gaussian simulation

- Plurigaussian simulation

Y. N. Pandey et al., *Machine Learning in the Oil and Gas Industry*, https://doi.org/10.1007/978-1-4842-6094-4_5

In this chapter, we discuss how machine learning can help us estimate petrophysical property values on a 3D grid. First, we discuss some of the basic foundations that are common to all the techniques, as well as the machine learning techniques that are used in this chapter. Following that, we describe the dataset that we use in this chapter, show application of machine learning algorithm on the dataset, and outline some algorithmic limitations.

Variogram

Variograms are the foundation of almost every 3D spatial modeling technique of industrial relevance. Variograms provide the semivariance (often called *variance* for simplicity) of property Z at different distances from the point of interest. In a classical geostatistical approach [1], the semivariogram $\gamma\,(h)$ (often called a *variogram*) of property Z is defined as

$$\gamma(h) = \frac{1}{2n(h)} \sum_{n(h)} E\left[\left(Z(u+h) - Z(u) \right)^2 \right], \qquad (5.1)$$

where $n(h)$ is the number of pairs that are separated by the distance h (also called *lag*). It should be noted that the experimental variograms calculated using this equation are fitted using an analytical expression, such as a spherical variogram model.

Figure 5-1 demonstrates the schematic representation of the spherical variogram model fit of a variogram $\gamma\,(h)$. In Figure 5-1, r represents the variogram range. This value indicates the separation from a given point over which the correlation between the property values vanishes (i.e., property values are decorrelated). Sill is represented by s, which is the value of semivariance achieved by the variogram asymptotically. Sill value can be conceived as the variance of the property values at two points, located at a distance more than the variogram range (r). Nugget variance, n, is an estimate of the uncertainty in the experimental or measured data.

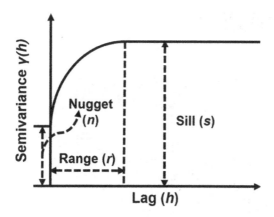

Figure 5-1. *Schematic depiction of a spherical variogram model*

The variogram may be omnidirectional or isotropic, where it is assumed that the same semivariance is observed for a lag value in any direction. It is also possible to have an anisotropic variogram, where semivariance depends on the lag value and the direction. An example of anisotropic variogram may be to use the same variogram model along the lateral directions (e.g., *x* and *y*), and use of a separate variogram model along the depth. Kriging techniques rely on variograms in the estimation of the petrophysical property value at a point away from the wells. Kriging is a form of generalized linear regression, which provides an optimal estimator to minimize mean-squared-errors of estimated property values. In this chapter, we are not going to discuss details of the kriging techniques. Rather, we focus on a machine learning algorithm, which is roughly equivalent to a simple kriging algorithm.

Having access to an open source data set for carrying out a 3D geomodeling exercise is the biggest challenge. For this chapter, our objective is to give you a working example of petrophysical properties interpolation over a 3D geological model. Keeping this in mind, we selected a simple geological model provided by the Society of Petroleum Engineers Comparative Solution Project SPE 9 [3], which is available at https://github.com/OPM/opm-data. The SPE 9 model is shown in Figure 5-2, with *permx* (permeability in the *x*-direction) shown in the overlay.

Data Description

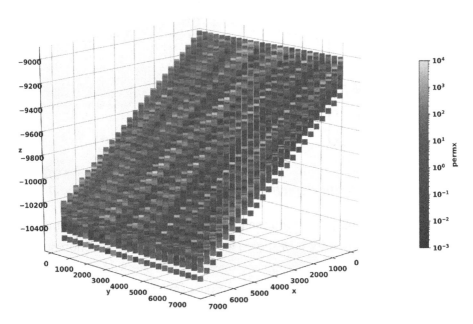

Figure 5-2. *The SPE 9 model is rendered with permx property as an overlay*

This grid has 24, 25, and 15 cells in the x, y, and z directions, respectively. By visually observing the data in this model, we can quickly note two clear features of this model. First, the model has a significant level of heterogeneity, and second, there is a low correlation between the values in the adjacent cells. The hypothesis formed is based on visual observation is examined by a quantitative analysis of the data in a later section of this chapter.

We extracted the well logs for the property *permx*, based on the well locations provided in the SPE 9 model, as listed in Table 5-1. Figure 5-3 shows the extracted well logs for the property *permx*.

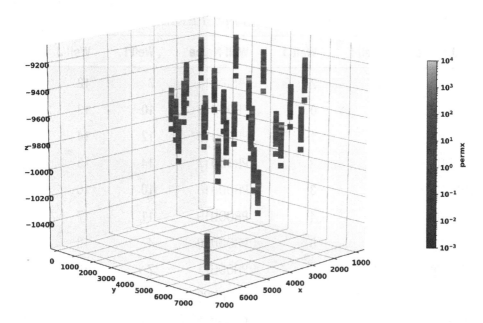

Figure 5-3. *Well logs for permx property, based on well locations for SPE 9 model*

In Figure 5-3, an isolated injector well can be seen in the northeast corner of the reservoir model. In contrast, producer wells are concentrated away from the injector well in a clustered manner.

Next, we perform some basic analysis of grid data and extracted well logs.

Table 5-1. *Well Location Indexes in the SPE 9 Model Grid*

Well Name	Well I	Well J	Well Name	Well I	Well J
INJE1	23	24	PRODU14	7	12
PRODU2	4	0	PRODU15	10	13
PRODU3	7	1	PRODU16	12	14
PRODU4	10	2	PRODU17	14	15
PRODU5	9	3	PRODU18	10	16
PRODU6	11	4	PRODU19	11	17
PRODU7	3	5	PRODU20	4	18
PRODU8	7	6	PRODU21	7	19
PRODU9	13	7	PRODU22	10	20
PRODU10	10	8	PRODU23	14	21
PRODU11	11	9	PRODU24	11	22
PRODU12	9	10	PRODU25	9	23
PRODU13	4	11	PRODU26	16	24

Basic Data Analysis of the Well Logs

We want to explore whether the *permx* data populated on the SPE 9
3D model, and the well logs extracted from this model follow the same
statistical patterns. Taking a pragmatic approach, we first calculate the
sample mean, standard deviation, and variance by using the *permx* values
from the extracted well logs, and all grid cells of the SPE 9 model. These
values are summarized in Table 5-2.

Table 5-2. *Mean and Standard Deviations of the permx Property for the Wells Extracted from the SPE 9 Model and for all Grid Cells*

Extracted Well Logs			All SPE 9 Grid Cells		
Mean	Standard Deviation	Variance	Mean	Standard Deviation	Variance
116.47	507.24	257287.53	108.08	375.31	140856.44

The values of mean *permx* property from the extracted well logs, and all the grid cells in the SPE 9 model are relatively close. However, the values of standard deviation and variance indicate that the distribution of values around the mean is significantly wider for the extracted well logs. A possible conclusion is that the *permx* property values on the SPE 9 model may not have been generated using the property distribution on the well locations provided with the model. We also compute experimental variogram models from the extracted well logs, and all the grid cells by using the code in Listing 5-1, which implements Equation 5.1 in Python.

Listing 5-1. Code Snippet for Experimental Variogram Computation

```python
from scipy.spatial.distance import pdist, squareform
import numpy as np
def variogram_one_lag(X, perm, lag, bin_width):
    """Experimental semi-variogram for one lag"""
    dist_mat = squareform(pdist(X))
    n_points = dist_mat.shape[0]
    list_var = []
    for i in range(n_points):
        for j in range(i + 1, n_points):
            if(dist_mat[i, j] >= (lag - bin_width)) and \
            (dist_mat[i, j] <= (lag + bin_width)):
                list_var.append((perm[i] - perm[j])**2.0)
    return np.sum(list_var) / (2.0 * len(list_var))
```

```python
def variogram(X, perm, lags, bin_width):
    """Experimental variogram calculation"""
    variogram = []
    for lag in lags:
        variogram.append(variogram_one_lag(X, perm, lag,
        bin_width))
    return variogram
```

Figure 5-4 shows the computed experimental variograms for the extracted well logs and the entire grid. The first fact that stands out is significantly large nugget variance in both of these variogram models. A large nugget indicates that the data is very noisy, and there is not enough spatial correlation.

This was a hypothesis, which we had from the visual observation, and now we can quantitatively establish this by using variogram computations. At this point it's clear that 3D interpolation based on property distribution at the well locations will not be able to reproduce the *permx* property model shown in the SPE 9 model (Figure 5-2). It may be tempting to argue that we can use an analytical variogram with zero nugget, but that is not realistic, as it does not capture the uncertainty inherent in the input data being provided for interpolation.

In the next section, we start discussing Gaussian process regression, which is a machine learning algorithm with parallels to kriging concepts, and we demonstrate how to use it for 3D interpolation of the extracted well logs over the SPE 9 grid cells.

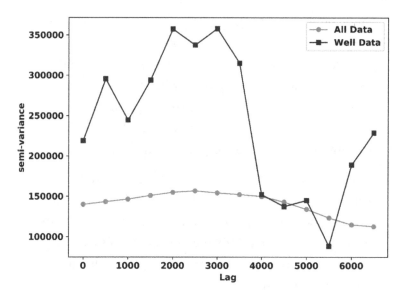

Figure 5-4. *Computed experimental variograms from the SPE 9 model, based on data from all grid cells (red circles), and data from the well logs extracted based on well locations (blue squares)*

Gaussian Process Regression

Gaussian process regression is a supervised learning method, which implements Gaussian processes to solve regression problems. There are certain advantages of Gaussian process regression, which make it appealing for 3D spatial interpolation problems.

- This algorithm has the excellent capability of interpolating the observed data. As a result, property values estimated in an area between already drilled wells are very reliable.

185

- Predictions generated using Gaussian process regression are probabilistic. Therefore, it is possible to compute confidence intervals and provide estimates of uncertainty on the estimated value. This functionality is implemented in the scikit-learn library. This is very important for geological modeling, as uncertainty quantification is a very important aspect in 3D property modeling

- Gaussian process regression uses the kernel trick. The kernels project problem input onto a high dimensional space, such that a highly nonlinear regression problem becomes linear, or approximately linear. This allows a more efficient solution to a challenging nonlinear regression problem.

However, just like any other algorithm, there are some disadvantages associated with Gaussian process regression as well. Some of the disadvantages include the following.

- The model training requires the entire sample information. This makes it unfit for training on very large datasets. However, geomodeling is a field where geoscientists must deal with a lack of data. Therefore, in most of the scenarios, this does not pose a serious challenge.

- Gaussian process regression works well if there are a limited number of variables in the input feature, approximately a few dozens. As we move to a higher dimensional input feature, the method loses its efficiency. This is also not a significant problem because, for spatial 3D interpolation, the input feature mostly consists of x, y, and z (or transformed u, v, and w) coordinates.

- This method is not suitable for extrapolation. As we move further away from the sample data, the estimated values tend to the mean of sample data, with estimation variance tending to the variance of the sample data.

Formulation

The Gaussian process (GP) is a collection of random variables, where any finite collection of those variables has a multivariate normal distribution. If we have a single random variable with the Gaussian (also normal) distribution, and we plot it, we get a bell-shaped curve. If we have two random variables with similar distributions, by plotting them on the x and y axes, we get the shape of a bell. That's a multivariate normal distribution with two random variables. We can imagine three random variables that satisfy the condition, and so on.

Now, let's say there are some points X, where observed property values are given by $f(X)$ or simply f. Also, based on this information, for a set of points X^*, we are trying to estimate $f(X^*)$ or f^*. This means that we are trying to calculate the probability distribution $p(f^*|f)$. Under the assumption of a zero mean, we can approximate $f(X)$ as following [4].

$$f(X) \sim N\big(0, K(\theta, X, X')\big), \qquad (5.2)$$

where N denotes a normal distribution with a zero mean, and $K(\theta, X, X')$ represents the covariance matrix between all observed data point pairs (X, X'). The covariance matrix ensures that the values in close proximity to the input generate output values, which are also close to each other. The estimated mean values at points X^* is given by

$$y = K\big(\theta, X^*, X\big) K\big(\theta, X, X'\big)^{-1} f(X). \qquad (5.3)$$

In Equation 5.3, a new term $K(\theta, X^*, X)$ appears. This term accounts for the covariance between the observed data points (in our case exported well logs), and the points where we are trying to get property estimates. If we draw a parallel between kriging and Gaussian process regression, the covariance matrix is similar to the analytical variogram shown in Figure 5-1. In Gaussian process regression, covariance function is implemented using the previously discussed *kernel trick*. The most commonly used kernel is the Radial-basis function (RBF) kernel.

Radial-Basis Function (RBF) Kernel

The RBF kernel is also known as the *squared exponential* kernel. This kernel is parameterized by length-scale parameters, which are similar to the variogram range shown in Figure 5-1. The RBF kernel can have a single length scale and provide an equivalent to omnidirectional (isotropic) variogram, or it may have the same number of dimensions as the input providing an equivalent to the anisotropic variogram. The functional form of the RBF kernel is given by

$$k(x,y,z) = \exp\left(-\frac{1}{2}d\left(\frac{x}{l_x}, \frac{y}{l_y}, \frac{z}{l_z}\right)^2\right). \tag{5.4}$$

In Equation 5.4 l_x, l_y, and l_z represent length-scales or variogram ranges in the x, y, and z directions. Also, the GaussianProcessRegressor scikit-learn implementation provides a hyperparameter, alpha, which represents the nugget variance (n) of the variogram model in Figure 5-1. It is important to note that the method discussed here assumes a zero sample mean. This can be achieved by using StandardScaler in scikit-learn. Listing 5-2 shows an implementation of the interpolation of *permx* using GaussianProcessRegressor.

Listing 5-2. Code Snippet for Gaussian Process Regression for Property Value Estimation Away from Wells

```
from sklearn.preprocessing import StandardScaler
from sklearn.gaussian_process import GaussianProcessRegressor
from sklearn.gaussian_process.kernels import RBF

x_scaler = StandardScaler()
X_scaled = x_scaler.fit_transform(X)
y_scaler = StandardScaler()
y_scaled = y_scaler.fit_transform(np.array(y).reshape(-1, 1))
kernel = RBF([5.e-1, 5.e-3, 5.e-3], (1.e-3, 1.e0))
gpr = GaussianProcessRegressor(kernel=kernel,
                               alpha=0.75,
                               n_restarts_optimizer=10,
                               normalize_y=False,
                               random_state=12345)
y_scaled = y_scaled.flatten()
gpr.fit(X_scaled, y_scaled)
pred, sigma = gpr.predict(X_scaled, return_std=True)
pred = y_scaler.inverse_transform(pred.reshape(1, -1))
print('GPR Fit Score: %f' % gpr.score(X_scaled, y_scaled))
```

GPR Fit Score: 0.708586653049732

In the code sample, we transformed the data to have zero mean and unit variance. Having unit variance also helps us choose the `alpha` parameter as a fraction of sample variance. A smaller value of `alpha` fits the data at the well locations extremely well, and we'll get a higher fit score. You can try experimenting with different values of `alpha` to see how the fit score and overall predictions change.

The RBF kernel in the code has three length scales, which indicates an anisotropic kernel (variogram) is being used. The parameters in the RBF kernel after the list of length scales is a tuple defining upper and lower bounds of the length-scales. During the training process, the optimization algorithms choose the best fitting length-scales within the defined upper and lower bounds. Using the model generated in Listing 5-2, values of *permx* are predicted at the well locations. The estimates have low variance, as shown in Figure 5-5.

Further, the *permx* values at all grid cells of the SPE 5 model are also predicted using the trained model. Figure 5-6 shows the estimated mean predictions and corresponding variances. It can be observed that the estimates around the wells have a low variance. As we move far away from the wells, we get predictions and variances, which are equal to the mean and variance values calculated from the exported well log data as shown in Table 5-2. We expected that this would happen based on the variogram models observed in Figure 5-4, where nugget variance was significant when compared to sill, indicating low correlation in the exported well log data.

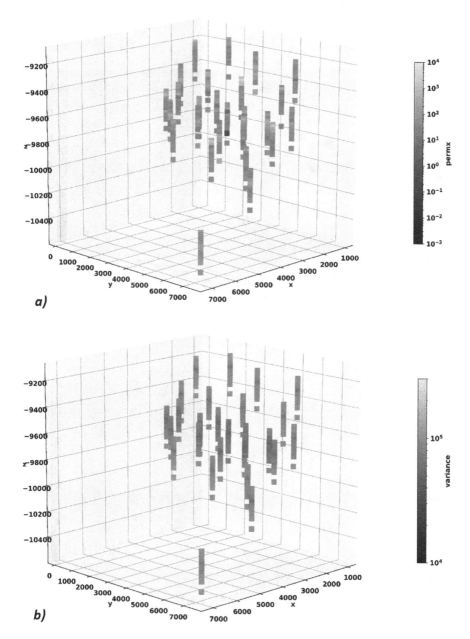

Figure 5-5. *Using Gaussian process regression (a) predicted permx values at the SPE 9 well locations and (b) corresponding estimation variance*

191

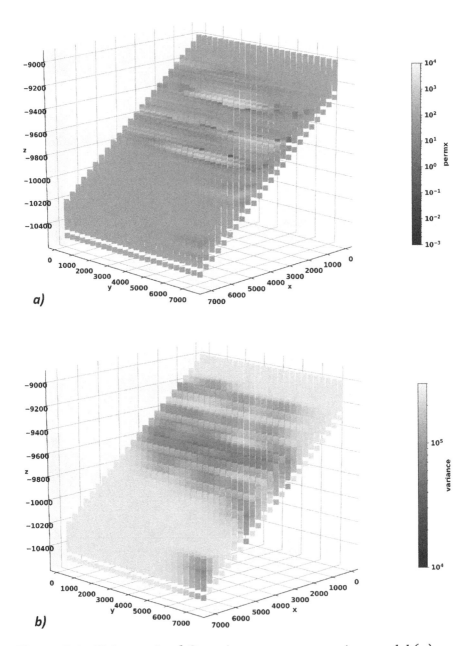

Figure 5-6. *Using trained Gaussian process regression model (a) predicted permx values at all SPE 9 grid cells and (b) corresponding estimation variance*

Summary

In this chapter, we used a machine learning method that provides a tool equivalent to the kriging technique. We used an open source data set to train the Gaussian process regression model. During recent years, there have been several new developments in the area of spatial interpolation. The most notable are *generative adversarial networks* (GANs). In the oil and gas industry, GANs have been used for the interpolation of seismic images [5]. However, it may be worthwhile to take a step back and assess the applicability of these methods in the geomodeling perspective.

Often, the biggest challenge facing the geoscientists is the availability of good data from the well logs. During the initial field development phase, the data from well logs is scarce. In such scenarios, the applicability of algorithms, such as GAN, is questionable given the large amounts of data required for training them. However, in the scenarios where well logs from hundreds of wells are available, it may be worthwhile to evaluate the applicability of deep learning-based techniques [6]. However, in a normal scenario, Gaussian process regression presents a good choice of algorithm for petrophysical properties interpolation.

EFFECT OF PARAMETER ALPHA

The results shown in this chapter were generated using an `alpha` parameter value of 0.75 while training the `GaussianProcessRegressor` model.

- Try using different values of the `alpha` parameter to see how it changes the output. You will see that the `GaussianProcessRegressor` score keeps improving as you lower the value of `alpha`, but its ability to predict property values away from the wells will diminish.

- If you have access to a realistic 3D grid model in .GRDECL
  format, try to use the code provided with the book for that 3D
  grid and see whether you can generate good results using a
  `GaussianProcessRegressor`.

References

[1] E. H. Isaaks and R. M. Srivastava, *An Introduction to Applied Geostatistics*, Oxford University Press, 1990.

[2] M. Armstrong, A. Galli, H. Beucher, G. Loc'h, D. Renard, B. Doligez, R. Eschard, and F. Geffroy, *Plurigaussian Simulations in Geosciences*, Berlin/Heidelberg: Springer-Verlag, 2011.

[3] J. Killough, "Ninth SPE Comparative Solution Project: A Reexamination of Black-Oil Simulation," SPE Reservoir Simulation Symposium, San Antonio, TX, 1995.

[4] C. E. Rasmussen and C. K. I. Williams, *Gaussian Processes for Machine Learning*, Cambridge: The MIT Press, 2006.

[5] D. A. B. Oliveira, R. S. Ferreira, R. Silva, and E. V. Brazil, "Interpolating Seismic Data With Conditional Generative Adversarial Networks," *IEEE Geoscience and Remote Sensing Letters,* vol. 15, no. 12, pp. 1952–1956, 2018.

[6] Y. N. Pandey, K. P. Rangarajan, J. M. Yarus, N. Chaudhary, N. Srinivasan, and J. Etienne, "Deep Learning Based Reservoir Modeling," Patent PCT/US2017/043228, July 21, 2017.

CHAPTER 6

Reservoir Engineering

The oil and gas industry has been solving problems related to automation and optimization from the very beginning. Modern technology and data-driven algorithms have provided the industry with an additional mechanism to solve problems and gain insights. Machine learning applications in the oilfield are observed in drilling engineering for ROP optimization [1], differential pipe sticking [2], identification of sweet spots [3], petrophysical modeling [4], reservoir fluid property estimation [5], subsurface characterization [6], reservoir simulation [7], production engineering [8], and completions engineering [9] to name a few.

Some studies have tried to combine traditional statistical methods with machine learning [10] to improve the performance of completion jobs and overall production. A hot topic in recent years is the optimization of completion design to mitigate fracture hits [11], which is found to be critical for unconventional plays [12]. Due to the existing heterogeneity and uncertainty in the reservoir, many have raised issues on pure physics-based models [13], which pushes our attention to data-driven solutions to complement those studies. Modern computational power has also enabled the applications of neural networks and deep learning-based solutions for solving problems in the oil and gas industry. Multiple studies have tried to use this learning technique to solve problems like bit-bounce detection [14], production data analysis [15], seismic horizon interpretation [16], and more.

© Yogendra Narayan Pandey, Ayush Rastogi, Sribharath Kainkaryam,
Srimoyee Bhattacharya, and Luigi Saputelli 2020
Y. N. Pandey et al., *Machine Learning in the Oil and Gas Industry*,
https://doi.org/10.1007/978-1-4842-6094-4_6

Applications of machine learning can be divided into two categories [12]. The first category relies on production data, and the second category includes completions, operations, and other reservoir parameters. This chapter is based on the first category, in which publicly available production data is used for the analysis.

In the field of petroleum engineering, one of the primary responsibilities of a reservoir engineer is the ability to quantify the reservoir's performance. This is performed using *decline curve analysis* (DCA), where the production rates are monitored against time, and predictions are made on oil or gas volumes, which can be produced by the reservoir. DCA plays an important role and is one of the most valuable techniques in a reservoir engineer's toolbox because it provides a way to estimate the initial hydrocarbon in place and the hydrocarbon reserves at the time of abandonment, and to forecast future production until economic limits are reached. Forecasts made using this approach are often used to report reserves to the US Securities and Exchange Commission (SEC), and hence they hold high importance in the field of petroleum economics.

With the increased adoption of machine learning-based solutions in the oil and gas industry, there has been an effort to improve decline curve analysis to make more accurate forecasts. One of the reasons behind the increasing application of machine learning is the massive amount of data, which helps build reservoir models and perform history match. The conventional methods of reservoir engineering and reservoir modeling are costly and time-consuming, hence leading our attention to data-driven methods that provide estimates with similar or improved accuracy.

This chapter introduces problems in the field of reservoir engineering, which can be solved using machine learning. Subsequent sections are dedicated to scraping a publicly available dataset and developing forecasting models for decline curve analysis.

Traditional forms of statistical analysis that use Arps decline methods and time series-based analysis to forecast production are introduced in this chapter. The forecasted results from different approaches are discussed. This is only a brief description, and the readers are encouraged to try other methods, like recurrent neural networks (RNNs) and long-short term memory (LSTM), to forecast oil and gas production, which have been shown to provide predictions with higher accuracy [17].

The Role of Machine Learning in Reservoir Engineering

Some of the areas of reservoir engineering in which machine learning-based solutions are directly applicable include optimized reservoir management [18], where identifying and planning of new wells is always a big challenge. Subsurface characterization of reservoir properties for an efficient field development plan can also be achieved by machine learning [5]. Another area where machine learning can be applied includes the generation of pseudo logs or prediction of PVT properties using data from analogous wells [19][20]. Some other examples include the interpretation of geology [21], geophysics [22], and the creation of knowledge graphs, which can provide deeper insights.

Literature has shown how certain machine learning–based methods like neural network and deep learning have improved the predictions in comparison to Arps, Duong and the *stretched exponential production decline* (SEPD) model [13] [23] [24]. By using machine learning algorithms, the traditional approach of using type wells for production forecasting [25] has been improved. Algorithms like random forest, support vector machine, and multivariate adaptive regression splines have also improved prediction accuracy [26]. Other studies have shown better results with neural networks for predicting oil rates for Bakken shale wells [27]. A typical decline curve with respect to time is shown in Figure 6-1.

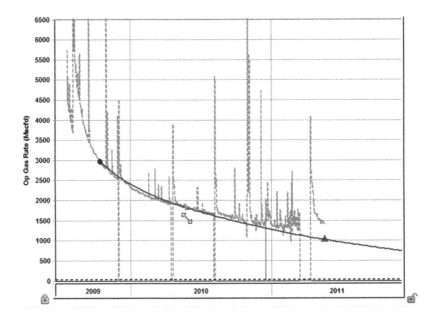

Figure 6-1. *An example of a typical decline curve with time on the x-axis and operating gas rates on the y-axis [28]*

Dataset Description

Publicly available production data from the Oil and Gas Division of the North Dakota Industrial Commission (NDIC) [29] was used in the analysis. A Python script scraped the data from the website. Excel (.csv) files from mid-2015 onward are available on the website, which were used in this chapter. A snapshot of the NDIC website is shown in Figure 6-2.

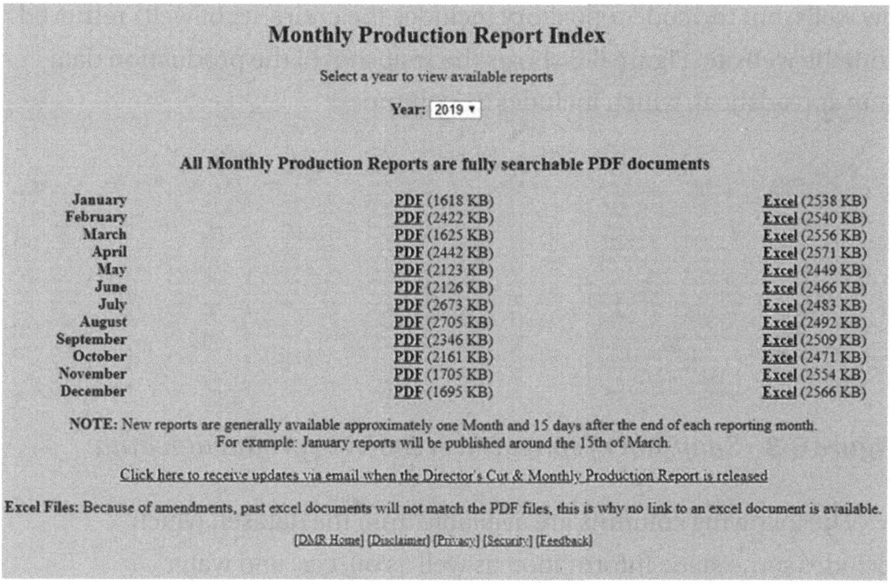

Figure 6-2. *Snapshot of North Dakota Industrial Commission website used to scrape production data [29]*

Once the data is scraped, as you can see in the distinct API numbers, there are 17,756 wells in the analysis. Not all the wells have oil and gas production reported accurately, and hence a challenge is to clean the data as much as possible. Using some basic data visualization, we can understand the dataset we are currently working with.

The analysis is focused on oil production, since it is the dominant flow from wells in North Dakota. In this study, we use data from mid-2015 through 2018 for history matching, and the data from 2019 is forecasted. The data is in the form of a time series with monthly production available for the analysis. Apart from the conventional Arps equations used for decline curve analysis, time series-based forecasting is also performed in this chapter.

Based on the two methods, we can determine the more accurate method using root-mean-square error (RMSE) as a metric, since this is a regression problem. It is to be noted that the analysis is only shown for a few wells, but the code repository includes the entire set of wells retrieved from the website. Figure 6-3 shows the snapshot of the production data from spreadsheet, which includes 21 columns.

	A	B	C	D	E	F	G	H	I	J	K	L	M	N	O	P	Q	R	S	T	U	V
1		ReportDate	API_WELLN	FileNo	Company	WellName	Quarter	Section	Township	Range	County	FieldName	Pool	Oil	Wtr	Days	Runs	Gas	GasSold	Flared	Lat	Long
2	0	5/1/2015	3.31E+13	23615	MUREX PETROL	AMBER ELL	SESW	36	151	101	MCK	ALEXANDE	BAKKEN	1525	1335	29	1686	1739	1519	0	47.84914	-103.553
3	1	5/1/2015	3.31E+13	22021	STATOIL OIL & G	BILL 14-23	SWSW	11	151	101	MCK	ALEXANDE	BAKKEN	1410	5805	31	1370	1721	1720	1	47.90743	-103.58
4	2	5/1/2015	3.31E+13	25091	STATOIL OIL & G	BILL 14-23	NWNE	14	151	101	MCK	ALEXANDE	BAKKEN	5035	12773	28	5087	3754	3249	505	47.90455	-103.569
5	3	5/1/2015	3.31E+13	25645	STATOIL OIL & G	BILL 14-23	NWNE	14	151	101	MCK	ALEXANDE	BAKKEN	2419	17671	31	2287	232	201	31	47.90455	-103.569
6	4	5/1/2015	3.31E+13	25092	STATOIL OIL & G	BILL 14-23	NWNE	14	151	101	MCK	ALEXANDE	BAKKEN	771	17718	31	729	53	46	7	47.90455	-103.569
7	5	5/1/2015	3.31E+13	25644	STATOIL OIL & G	BILL 14-23	NWNE	14	151	101	MCK	ALEXANDE	BAKKEN	5558	11971	31	5316	5196	4497	699	47.90455	-103.569
8	6	5/1/2015	3.31E+13	22023	STATOIL OIL & G	BILL 14-23	SWSW	11	151	101	MCK	ALEXANDE	BAKKEN	6848	6926	31	6748	8537	7791	746	47.90748	-103.58
9	7	5/1/2015	3.31E+13	23319	STATOIL OIL & G	PORTER 3	SESW	35	151	101	MCK	ALEXANDE	BAKKEN	1763	5903	31	1708	2141	2140	1	47.84919	-103.577
10	8	5/1/2015	3.31E+13	23320	STATOIL OIL & G	PORTER 3	SESW	35	151	101	MCK	ALEXANDE	BAKKEN	1973	3258	31	1925	2140	2139	1	47.84919	-103.577
11	9	5/1/2015	3.31E+13	23302	STATOIL OIL & G	TIMBER CR	NENW	13	151	101	MCK	ALEXANDE	BAKKEN	3123	6513	30	3200	3159	2580	579	47.90519	-103.551
12	10	5/1/2015	3.31E+13	23301	STATOIL OIL & G	TIMBER CR	NENW	13	151	101	MCK	ALEXANDE	BAKKEN	7669	7460	31	8009	8792	8638	154	47.90519	-103.551
13	11	5/1/2015	3.31E+13	9132	TAQA NORTH U:	SKEDSVOLL	SENW	23	151	101	MCK	ALEXANDE	DUPEROW	226	3	11	223	59	0	0	47.88669	-103.575
14	12	5/1/2015	3.31E+13	9617	TAQA NORTH U:	BOLKEN	NWSW	24	151	101	MCK	ALEXANDE	MADISON	0	0	0	0	0	0	0	47.88262	-103.559
15	13	5/1/2015	3.31E+13	8165	TAQA NORTH U:	SKEDSVOLL	SENE	23	151	101	MCK	ALEXANDE	MADISON	201	1392	22	337	93	0	0	47.88621	-103.565
16	14	5/1/2015	3.31E+13	9844	TAQA NORTH U:	SOVIG 2:	NWNE	25	151	101	MCK	ALEXANDE	MADISON	0	0	0	0	0	0	0	47.87541	-103.548
17	15	5/1/2015	3.31E+13	15670	TAQA NORTH U:	HANSEN 2	W2SE	23	151	101	MCK	ALEXANDE	RED RIVER	0	0	0	0	0	0	0	47.88086	-103.569
18	16	5/1/2015	3.31E+13	8165	TAQA NORTH U:	SKEDSVOLL	SENE	23	151	101	MCK	ALEXANDE	RED RIVER	53	0	22	90	57	0	0	47.88621	-103.565
19	17	5/1/2015	3.3E+13	23323	HUNT OIL COMF	ALEXANDR	LOT3	1	160	101	DIV	ALEXANDR	BAKKEN	1147	1854	28	1009	1247	0	584	48.71889	-103.678
20	18	5/1/2015	3.3E+13	28507	HUNT OIL COMF	ALEXANDR	SWSE	21	161	100	DIV	ALEXANDR	BAKKEN	6386	8133	31	6129	3663	0	3370	48.75011	-103.668

Figure 6-3. *Snapshot of spreadsheet showing production data*

The following columns are available from the dataset, which includes some static information as well as oil, gas, and water production reported on a monthly basis. The location of the wells in the form of latitude and longitude is also provided in the database. Information about the number of production days determines the oil rate.

The following are the 21 data columns.

```
Index(['ReportDate', 'API_WELLNO', 'FileNo', 'Company',
'WellName', 'Quarter', 'Section', 'Township', 'Range',
'County', 'FieldName', 'Pool', 'Oil', 'Wtr', 'Days', 'Runs',
'Gas', 'GasSold', 'Flared', 'Lat', 'Long'], dtype="object")
```

Some of the additional features added to the existing dataframe include the date when the well came online, as well as finding the number of days for which the particular well was online. The following code snippet finds the top ten wells with the highest oil production in the database.

```
# find the top 10 wells with highest production (sum)
grouped_data = train_prod.groupby(['API_WELLNO']).sum()
grouped_data = grouped_data.sort_values(by=['Oil'])
grouped_data = grouped_data.nlargest(10, 'Oil').reset_index()
```

Decline trends for two example wells are shown in Figure 6-4 with oil production rates on the *y*-axis and time on the *x*-axis.

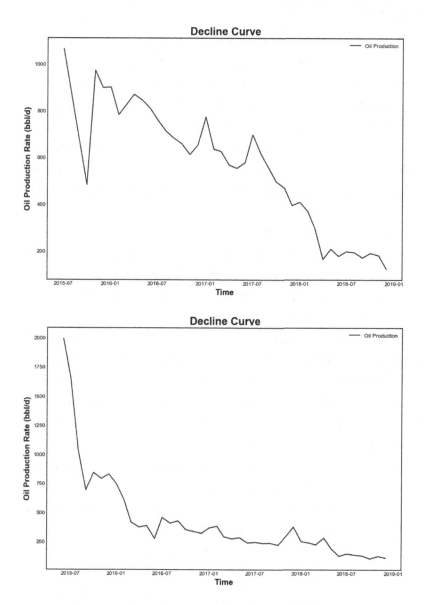

Figure 6-4. *Decline curve of two example wells (oil production rates plotted against time scale)*

Decline Curve Analysis Using the Arps Empirical Model

In this section, we focus on a decline curve analysis, which is a technique used by reservoir engineers to generate a forecast of future production rates of an oil or gas well. The results then determine the *estimated ultimate recovery* (EUR). The most commonly used equations are in the form of the empirical relationship between decline parameters, rate, and time as provided in a previous study [29], and are still commonly used. An integral component of modern production data analysis, the process also includes flow regime characterization, recovery factor estimation, and estimation of fluid-in-place.

The equations for decline curve analysis using this method are shown in the following section. This traditional graphical approach does not have a theoretical basis, and implicitly assumes constant operating conditions for the life of the well. Arps model has been successfully used in wells that have reached *boundary dominated flow* (BDF).

A note of caution regarding the majority of unconventional wells that have not reached boundary dominated flow and show steep declines in the transient period itself: This approach does not use pressure data either. An example of a well is shown where a curve is fit using parameters for *qi*, *b*, and *Di,* and a hyperbolic equation is used. A forecast is made for the 2019 period using the fitted curve.

To match historical data and make predictions, SciPy, a Python-based open source library, is used. SciPy provides a wrapper function— `curve_fit()` around `scipy.optimize.leastsq` [30]. This function uses a nonlinear least-squares method to fit a function and determine the necessary coefficients. The `curve_fit` method takes in three parameters at least.

$$popt, pcov = curve\_\text{fit}\,(func,\, T,\, Q).\qquad(6.1)$$

The first argument in the function call is a function. Any function can be passed to it as if it's any other variable. The function in our case is exponential, hyperbolic, or harmonic, and can take independent variable (e.g., time for the decline curve example) as well as any other fitting parameter. For example, in a hyperbolic function, qi, b, and Di are passed, whereas in an exponential function, only qi and Di are required since $b=0$. Apart from the function definition, independent variable and observed data is also passed, which is required to be fit. An L2 norm can determine the quality of the fit where the actual value is compared to the curve-fitting prediction.

$$Error\,L2 = \sum |(t) - qobs(t)|^2 \qquad (6.2)$$

Equations and Descriptions

Each type of functional form used in decline curve analysis has specific features. Here we discuss some of the salient features for different types of decline curves.

Exponential Decline

In this type of decline, a straight line is observed between flow rate and cumulative production, as well as between the logarithms of rate against time on the x-axis. The decline rate is constant. Predictions from this equation yield the minimum *estimated ultimate recovery* (EUR), and the rate can be calculated by using the following equation.

$$q = \frac{q_i}{e^{D_i t}}. \qquad (6.3)$$

Hyperbolic Decline

This is the most common Arps equation since straight-line plots are rarely encountered in real-life scenarios. The decline rate is not constant while the decline exponent factor b lies between 0 and 1 for conventional reservoirs (cases in which boundary dominated flow is the dominant flow regime), and is often determined by nonlinear curve fitting approach. Higher values of b factor have been observed in unconventional reservoirs. An example of such a case is a transient linear flow, where $b=2$. The rate for reservoirs exhibiting hyperbolic decline can be obtained by using the following equation.

$$q = \frac{q_i}{\left(1+bD_it\right)^{\frac{1}{b}}}. \qquad (6.4)$$

Harmonic Decline

This is considered to be a special case of hyperbolic decline with a $b=1$. In this case, the decline rate is directly proportional to the flow rate. A semi-log plot with a logarithm of flow rate against cumulative production shows a straight line. The flow rate can be calculated using the following equation.

$$q = \frac{q_i}{1+D_it}. \qquad (6.5)$$

Results

History matching for the wells was performed, and the results are shown in Figure 6-5.

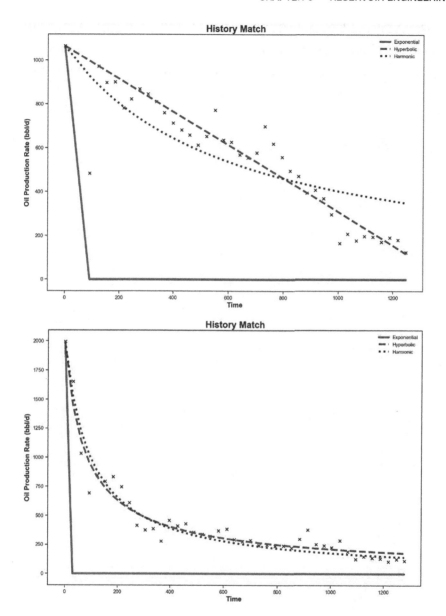

Figure 6-5. *History match results using Arps decline curve model*

For both example wells, the hyperbolic equation provides a better match. Although for the second well, hyperbolic and harmonic equations both show similar history match.

205

Using the model parameters, forecasts for oil production rates could be made by changing the time, t, in the equations. Figure 6-6 shows the results for forecasts made for the two wells for an additional 5000 days.

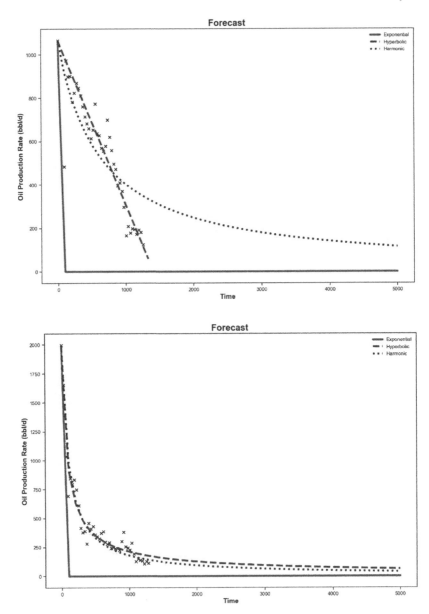

Figure 6-6. *Forecasted results using Arps decline curve model*

Validation

A test dataset which includes production from 2019 onwards is used to validate the performance of forecasts from the decline curve model. This data has not been used to train the empirical models and serves the purpose of validating the performance of the model. Figure 6-7 includes the comparison of hyperbolic curve performance with the actual data.

A common accuracy metric—root-mean-squared error (RMSE)—is determined in this example to quantify the accuracy of predictions from the hyperbolic model.

```
RMSE - Hyperbolic Model: 81.419 bbl/d
```

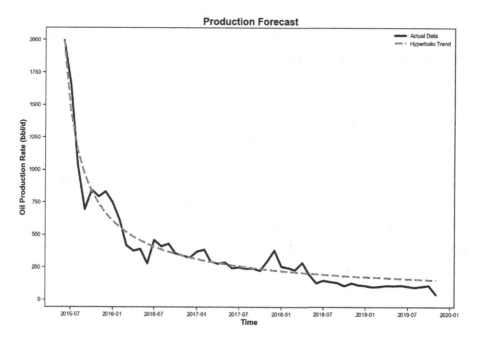

Figure 6-7. *Comparison between actual data and hyperbolic forecast which includes the test data (i.e., production from 2019 onward)*

Time Series–Based Forecasting Using the ARIMA Model

Another method for understanding the trend of the time series data and forecasting includes analyzing statistical trends from historical data. A common model, ARIMA (*autoregressive integrated moving average*), is used in this chapter. In this particular analysis, the date column needs to be converted to datetime datatype for easier manipulation of time data.

Stationarity Requirements for the Time Series

While working with time series modeling algorithms, we need to ensure the data is stationary. A stationary time series is one whose statistical properties, such as mean, variance, and autocorrelation, are constant over time [31]. By using mathematical transformations, such as logarithms, it is possible to convert the data into the form of stationary time series. Dickey-Fuller tests can be used to find out whether a time series is stationary or not [32].

The decline curve in the example for the well with API: 33025021780000 shows the presence of a time trend (usually decreasing for decline curve analysis), and seasonality, which makes the local mean a function of time. This proves the nonstationary nature of the decline curve, as shown in Figure 6-8. The exponentially weighted moving average approach is used to remove the time trend from the data. When the transformation is made, the data loses the trend and exhibits a close to the constant rolling mean.

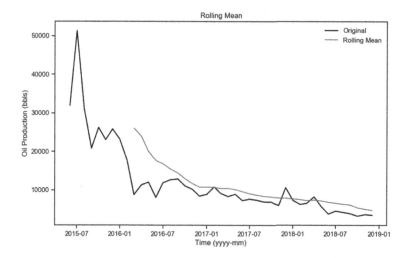

Figure 6-8. *Oil production data showing the decreasing trend with variable rolling mean indicating nonstationary behavior*

By using the method of differences, both seasonality and trends can be removed from the data. In this method, a difference of observation at a particular time is taken with the preceding observation. First-order differencing can be carried out in Python using the pandas library and is an effective approach in removing seasonality. The implementation is shown in Figure 6-9, which shows the loss of trend and seasonality after the transformations are applied.

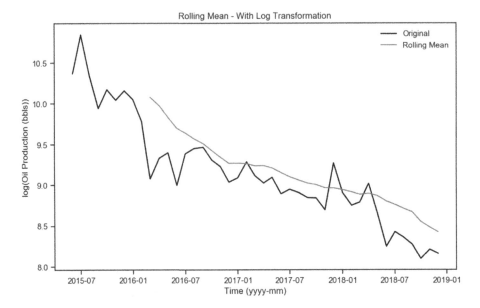

Figure 6-9. *Trend and seasonality removed using log transformation, leading to locally stable mean*

Statistical correlation analysis is performed to understand the strength of the relationship between two variables. This leads us to *autocorrelation function* (ACF) and *partial autocorrelation function* (PACF). ACF is defined as the coefficient of correlation between two values in a time series, which can measure the linear relationship between an observation at the current time, *t*, and the observations at previous times [33]. If the time series is transformed and the correlation coefficient is obtained, it gives us PACF, which is often useful for identifying the order of an autoregressive model. For the example discussed, ACF and PACF plots are shown in Figure 6-10.

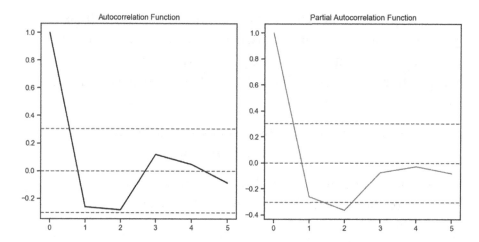

Figure 6-10. *ACF and PACF plots to understand the correlation between variables*

ARIMA Components

If you break down the *autoregressive integrated moving average,* or ARIMA, modeling algorithm, each component of the acronym has its own definition. ARIMA models time series data for forecasting (i.e., for predicting future points in the series), in such a way that the following occurs [34] [36].

- A pattern of growth/decline in the data is accounted for (hence the *autoregressive* component)

- The rate of change of the growth/decline in the data is accounted for (hence the *integrated* component)

- Noise between consecutive time points is accounted for (hence the *moving average* component)

211

Each of these components is explicitly specified in the model as a parameter. Standard notation is used in ARIMA (p, d, q), where the parameters are substituted with integer values to quickly indicate the specific ARIMA model being used [35] [37]. The parameters of the ARIMA model are defined as follows.

- *p*. The number of lag observations included in the model, which is also called the *lag order*

- *d*. The number of times that the raw observations are differenced, which is also called the *degree of differencing*

- *q*. The size of the moving average window, which is also called the *order of moving average*

The autocorrelation and partial autocorrelation functions (ACF and PACF) help determine the starting values of these parameters. By observing Figure 6-10, we can observe a sharp decline in the ACF and PACF after a certain number of lags (shown on the x-axis). By analyzing the lag for this decline, we can estimate the parameters of an ARIMA model. The parameter, p, for an autoregressive part, is determined using the PACF, whereas ACF provides the parameter q for the moving average part of the model. Based on the observations made in Figure 6-10, we are going to use an ARIMA model with the parameters $p=2$, $d=1$, and $q=2$. Next, we show plots for a residual sum of squares (RSS) obtained from the autoregressive model (see Figure 6-11), the moving average model (see Figure 6-12), and the combined ARIMA model (see Figure 6-13).

The Autoregressive Model

```
model_AR = ARIMA(ts_log, order=(2, 1, 0))
```

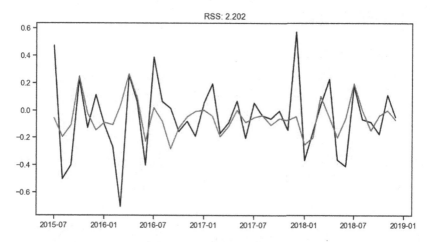

Figure 6-11. *Autoregressive model with residual sum of squares score of 2.202*

The Moving Average Model

```
model_MA = ARIMA(ts_log, order=(0, 1, 2))
```

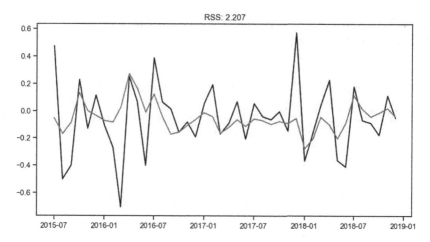

Figure 6-12. *Moving average model with a residual sum of squares score of 2.207*

213

The ARIMA Model

```
model = ARIMA(ts_log, order=(2, 1, 2))
```

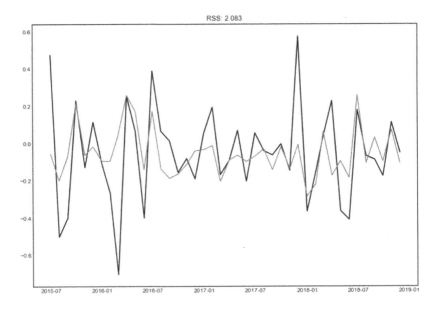

Figure 6-13. *ARIMA model with log transformation and 2.083 residual sum of squares*

Using the statsmodels library in Python, a summary of results can be generated, as shown in Figure 6-14.

```
                          ARIMA Model Results
===============================================================================
Dep. Variable:               D.Oil   No. Observations:                   42
Model:                ARIMA(2, 1, 2)  Log Likelihood                   4.240
Method:                     css-mle  S.D. of innovations              0.213
Date:            Wed, 08 Jul 2020   AIC                              3.520
Time:                    16:34:41   BIC                             13.946
Sample:                  07-01-2015  HQIC                             7.342
                       - 12-01-2018
===============================================================================
                 coef    std err          z      P>|z|      [0.025      0.975]
-------------------------------------------------------------------------------
const          -0.0552     0.015     -3.782      0.000     -0.084      -0.027
ar.L1.D.Oil    -0.6431     0.290     -2.216      0.027     -1.212      -0.074
ar.L2.D.Oil     0.0626     0.279      0.224      0.822     -0.484       0.609
ma.L1.D.Oil     0.3292     0.273      1.205      0.228     -0.206       0.864
ma.L2.D.Oil    -0.6707     0.255     -2.635      0.008     -1.170      -0.172
                                 Roots
===============================================================================
                  Real         Imaginary          Modulus         Frequency
-------------------------------------------------------------------------------
AR.1           -1.3718          +0.0000j           1.3718            0.5000
AR.2           11.6453          +0.0000j          11.6453            0.0000
MA.1           -1.0001          +0.0000j           1.0001            0.5000
MA.2            1.4909          +0.0000j           1.4909            0.0000
-------------------------------------------------------------------------------
```

Figure 6-14. *Summary of results from the ARIMA model*

It is difficult to visualize the results from the table, and hence a deeper analysis of residual and *kernel density estimation* (KDE) is performed. The results from the analysis of residual and KDE are shown in Figure 6-15 and Figure 6-16, respectively. In the residual plot, most of the residual points lie around the value of 0. The KDE plot shows the Gaussian distribution of the residuals. These plots indicate a good performance from the predictions.

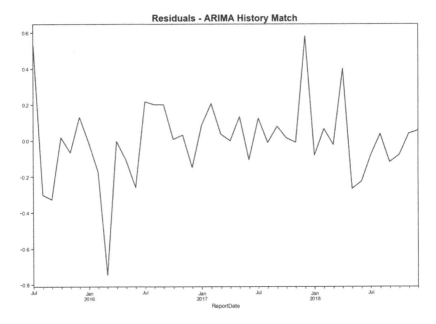

Figure 6-15. *Residual plot*

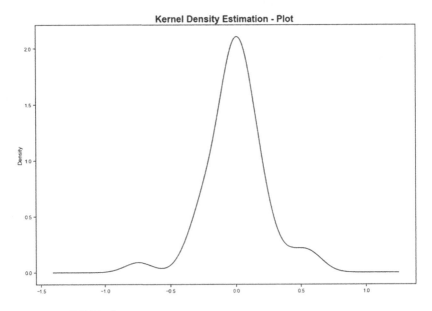

Figure 6-16. *KDE plot*

History match results from time series based ARIMA model are made for the oil production, and forecast for an additional year is shown in Figure 6-17. The plot also shows the window for a 95% confidence interval.

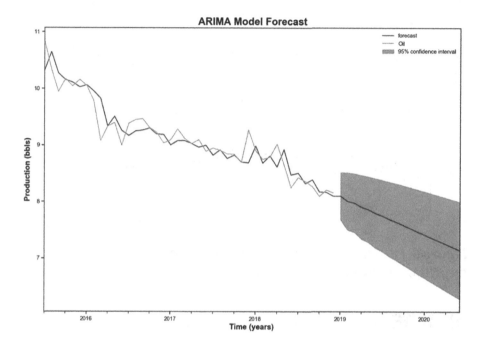

Figure 6-17. *History match and forecast for production data using the ARIMA model*

Similar to the process of using the Arps decline curve, data from 2019 was used as test data. Forecast results from the ARIMA model are compared to the actual values, with RMSE as an accuracy metric using the following code snippet. The validation curve is shown in Figure 6-18.

```
from sklearn.metrics import mean_squared_error
from math import sqrt
rmse = sqrt(mean_squared_error(actual, forecast))
print("RMSE - ARIMA Method:", rmse)

RMSE - ARIMA Method: 4141.44 bbls
```

217

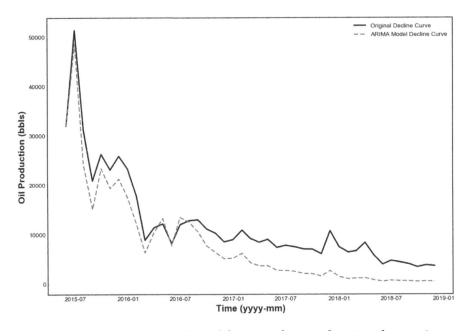

Figure 6-18. *History match and forecast for production data using the ARIMA model*

Summary

In this chapter, two different methods for production forecasting, namely Arps and ARIMA, were discussed. Production data from a well in North Dakota was used as an example; the information was scraped from a public website. Decline curve analysis using the Arps method is the conventional form of forecasting used in the field of reservoir engineering. The time series–based approach using ARIMA is statistically oriented and provides a machine learning flavor to the production forecasting problem. The application of these two methods shows their effectiveness in solving the problem. We encourage you to apply these methods in your day-to-day production forecasting.

References

[1] C. I. Noshi, "Application of Data Science and Machine Learning Algorithms for ROP Optimization in West Texas: Turning Data into Knowledge," Offshore Technology Conference, Houston, TX, 2019.

[2] A. Marana, J. Papa, M. V. Ferreira, K. Miura, and F. A. Torres, "An Intelligent System To Detect Drilling Problems Through Drilled-Cuttings-Return Analysis," IADC/SPE Drilling Conference and Exhibition, New Orleans, LA, 2010.

[3] S. Tandon, "Integrating Machine Learning in Identifying Sweet Spots in Unconventional Formations," SPE Western Regional Meeting, San Jose, CA, 2019.

[4] S. Vallabhaneni, R. Saraf, and S. Priyadarshy, "Machine-Learning-Based Petrophysical Property Modeling," SPE Europec at 81st EAGE Conference and Exhibition, London, England, UK, 2019.

[5] C. Onwuchekwa, "Application of Machine Learning Ideas to Reservoir Fluid Properties Estimation," SPE Nigeria Annual International Conference and Exhibition, Lagos, Nigeria, 2018.

[6] F. A. Anifowose, "Ensemble Machine Learning: The Latest Development in Computational Intelligence for Petroleum Reservoir Characterization," SPE Saudi Arabia Section Technical Symposium and Exhibition, Al-Khobar, Saudi Arabia, 2013.

[7] V. Gaganis and N. Varotsis, "Machine Learning Methods to Speed up Compositional Reservoir Simulation," SPE Europec/EAGE Annual Conference, Copenhagen, Denmark, 2012.

[8] M. Pennel, J. Hsiung and V. Putcha, "Detecting Failures and Optimizing Performance in Artificial Lift Using Machine Learning Models," SPE Western Regional Meeting, Garden Grove, CA, 2018.

[9] D. Castiniera, R. Toronyi, and N. Saleri, "Machine Learning and Natural Language Processing for Automated Analysis of Drilling and Completion Data," SPE Kingdom of Saudi Arabia Annual Technical Symposium and Exhibition, Dammam, Saudi Arabia, 2018.

[10] A. Rastogi and A. Sharma, "Quantifying the Impact of Fracturing Chemicals on Production Performance Using Machine Learning," SPE Liquids-Rich Basins Conference—North America, Odessa, TX, 2019.

[11] D. Fu, "Unlocking Unconventional Reservoirs With Data Analytics, Machine Learning, and Artificial Intelligence," *Journal of Petroleum Technology*, pp. 14–15, January 2019.

[12] S. Mohaghegh, "How Does the Use of Artificial Intelligence and Machine Learning Differ for Conventional vs. Unconventional Plays?," *Data Science and Digital Engineering in Upstream Oil and Gas*, October 7, 2019.

[13] Q. Cao, R. Banerjee, S. Gupta, J. Li, W. Zhou, and B. Jeyachandra, "Data Driven Production Forecasting Using Machine Learning," SPE Argentina Exploration and Production of Unconventional Resources Symposium, Buenos Aires, Argentina, 2016.

[14] M. Vassallo, G. Bernasconi, and V. Rampa, "Bit Bounce Detection Using Neural Networks," SEG Annual Meeting, Denver, CO, 2004.

[15] A. Rastogi, K. Agarwal, E. Lolon, M. Mayerhofer, and O. Oduba, "Demystifying Data-Driven Neural Networks for Multivariate Production Analysis," Unconventional Resources Technology Conference, Denver, CO, 2019.

[16] J. Lowell and G. Paton, "Application of Deep Learning for Seismic Horizon Interpretation," SEG International Exposition and Annual Meeting, Anaheim, CA, 2018.

[17] J. Sun, X. Ma and M. Kazi, "Comparison of Decline Curve Analysis DCA with Recursive Neural Networks RNN for Production Forecast of Multiple Wells," SPE Western Regional Meeting, Garden Grove, CA, 2018.

[18] V. Elichev, A. Bilogan, K. Litvinenko, R. Khabibullin, A. Alferov, and A. Vodopyan, "Understanding Well Events with Machine Learning," SPE Russian Petroleum Technology Conference, Moscow, Russia, 2019.

[19] M. Aliahmadi and Z. Chen, "Comparison of Machine Learning Methods for Estimating Permeability and Porosity of Oil Reservoirs via Petrophysical Logs," *Petroleum*, vol. 5, no. 3, pp. 271–284, 2019.

[20] R. Gharbi and A. M. Elsharkawi, "Neural Network Model for Estimating the PVT Properties of Middle East Crude Oils," *SPE Reservoir Evaluation & Engineering,* vol. 2, no. 03, 1999.

[21] T. Jobe and E. Vital-Brazil, "Geological Feature Prediction Using Image-Based Machine Learning," *Petrophysics,* vol. 59, no. 06, 2018.

[22] H. Maniar, S. Ryali, M. Kulkarni, and A. Abubakar, "Machine-learning methods in geoscience," SEG International Exposition and Annual Meeting, Anaheim, CA, 2018.

[23] D. Han, S. Kwon, H. Son, and J. Lee, "Production Forecasting for Shale Gas Well in Transient Flow Using Machine Learning and Decline Curve Analysis," SPE/AAPG/SEG Asia Pacific Unconventional Resources Technology Conference, Brisbane, Australia, 2019.

[24] Y. Li and Y. Han, "Decline Curve Analysis for Production Forecasting Based on Machine Learning," SPE Symposium: Production Enhancement and Cost Optimisation, Kuala Lumpur, Malaysia, 2017.

[25] A. Rastogi and W. J. Lee, "Methodology for Construction of Type Wells for Production Forecasting in Unconventional Reservoirs," Unconventional Resources Technology Conference, San Antonio, TX, 2015.

[26] A. Vyas, A. Datta-Gupta, and S. Mishra, "Modeling Early Time Rate Decline in Unconventional Reservoirs Using Machine Learning Techniques," Abu Dhabi International Petroleum Exhibition & Conference, Abu Dhabi, UAE, 2017.

[27] A. Suhag, R. Ranjith, and F. Aminzadeh, "Comparison of Shale Oil Production Forecasting using Empirical Methods and Artificial Neural Networks," SPE Annual Technical Conference and Exhibition, San Antonio, TX, 2017.

[28] "Traditional Decline Analysis Theory," [Online]. Available: http://www.fekete.com/SAN/WebHelp/FeketeHarmony/Harmony_ WebHelp/Content/HTML_Files/Reference_Material/Analysis_Method_ Theory/Traditional_Decline_Theory.htm.

[29] "North Dakota Oil and Gas Division," [Online]. Available: https:// www.dmr.nd.gov/oilgas/.

[30] J. Arps, "Analysis of Decline Curves," *Transactions of the AIME,* vol. 160, no. 01, 1945.

[31] "NumPY and Scipy Documentation," [Online]. Available: `https://docs.scipy.org/doc/`.

[32] B. Nau, "Stationarity and Differencing," [Online]. Available: `https://people.duke.edu/~rnau/411diff.htm`.

[33] J. Brownlee, "How to Check if Time Series Data is Stationary with Python," December 2016. [Online]. Available: `https://machinelearningmastery.com/time-series-data-stationary-python/`.

[34] I. Pardoe, "Autocorrelation and Time Series Methods," [Online]. Available: `https://online.stat.psu.edu/stat462/node/188/`.

[35] D. Abugaber, "Using ARIMA for Time Series Analysis," [Online]. Available: `https://ademos.people.uic.edu/Chapter23.html`.

[36] J. Brownlee, "How to Create an ARIMA Model for Time Series Forecasting in Python," [Online]. Available: `https://machinelearningmastery.com/arima-for-time-series-forecasting-with-python/`.

[37] K. Perry, "Time Series Forecasting Using a Seasonal ARIMA Model: A Python Tutorial," [Online]. Available: `https://techrando.com/2020/01/04/time-series-forecasting-using-a-seasonal-arima-model/`.

CHAPTER 7

Production Engineering

Production engineering deals with the analysis of field data and models to make decisions on how to optimally maintain the well and surface production facilities' performance. This chapter covers selected topics of production modeling using machine learning methodologies. The topics include model identification for predicting well rates, identification of early actions to achieve optimal production rates, and identification of the producing wells, which may benefit from the workover activities.

Production Engineering Overview

Production optimization encompasses various activities, including measuring, analyzing, modeling, prioritizing, and implementing actions to enhance productivity and profitability. Production engineers broadly consider multiple options to improve production system conditions technically, economically, and environmentally. As an example of an upstream integrated production system from the source reservoir to the point of sales, Figure 7-1 shows the schematic of a coalbed gas production system from the reservoir interface up to the sales point [1].

© Yogendra Narayan Pandey, Ayush Rastogi, Sribharath Kainkaryam,
Srimoyee Bhattacharya, and Luigi Saputelli 2020
Y. N. Pandey et al., *Machine Learning in the Oil and Gas Industry*,
https://doi.org/10.1007/978-1-4842-6094-4_7

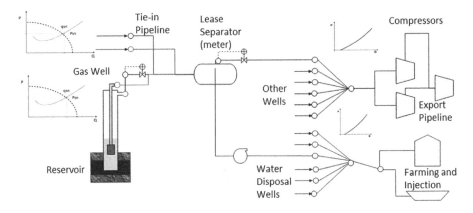

Figure 7-1. *Integrated gas-water production system [1]*

The relationship of flowing pressure as a function of liquid rate describes the steady-state well performance at any point in time (Figure 7-2). The outcomes of well optimization may be to enhance productivity in the reservoir (left), which results in higher liquid rates at higher flowing pressure and/or to reduce the restrictions in the outflow (right), which increases liquid rates at lower flowing pressures.

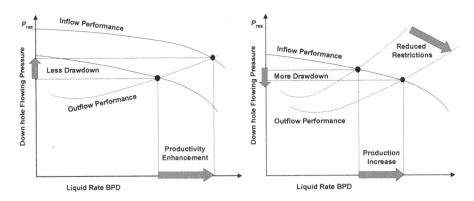

Figure 7-2. *Rate increase along inflow and outflow performance curves*

With the introduction of digital automation over the last couple of decades, a mature technology framework exists for enabling both one-off and continuous oil and gas field optimization [2]. Continuous optimization requires the integration of field hardware (e.g., downhole sensors, remotely activated completions, surface facilities) and computer algorithms enabling decision making, including data processing, virtual sensing, pattern recognition, and physics model integration.

Production engineering typically addresses the following questions through data analysis and physics modeling.

- Is the asset (reservoir, well, facility) producing up to its full potential?

- What is the surveillance data revealing about asset health conditions?

- What is the production limiting factor?

- What is the most likely cause of production loss/ deferral?

- What is the most effective action to arrest decline, and restore or increase production?

- If an action is taken, would it be successful, profitable, and sustainable?

Production engineering is a continuous process for enhancing technical and financial performance in the oil and gas fields. The goal is to maximize the effective production system capacity at the minimum cost and effort. This cyclic process involves routine data acquisition, analyze data and integrate information, options analysis and decision, and detailed engineering and execution (see Figure 7-3).

Routine Data Acquisition	**Analyze Data and Integrate Information**	**Options Analysis & Decision**	**Detailed Engineering and Execution**
Perform Site Inspections	Assess reservoir, well and facility status, rates and pressures	Specify modeling requirements	Plan surveillance
Reconcile daily rates and allocate monthly production	Model individual components and Integrated production system performance	Specify data requirements	Update models
Perform well flow an Pressure tests		Adjust well and facilities operating point	Engineering, planning and procurement
Install, maintain and leverage permanent monitoring systems	Establish capacity of isolated components and integrated system	Assess solution probabiity of success and sustainability	Site operations nad field work
	Identify Abnormal Events	Specify well work requirements	

Figure 7-3. *Production engineering key processes*

A large variety and volume of data are captured through routine surveillance programs, which are assimilated to identify abnormal events by exception. Accordingly, production engineers analyze and integrate all acquired data for assessing the status, modeling system performance and establishing the optimum operating envelope of wells and facilities. Such analysis allows tagging of key performance labels for well and facilities.

- Based on flowing status: active healthy, active unhealthy, and inactive

- Based on expected performance: as expected, above or below expected performance

- Based on problem identification: none, precursor event, problem, failure

- Based on potential and constrained capacity: reservoir, wells, facilities, legal, planned maintenance

With the introduction of machine learning to the oil and gas industry, numerous applications have been published to aid in the research of flowing status identification, performance metering, and problem identification and production capacity constraints.

Multiple methods are used to look for opportunities at the reservoir level, including drilling new wells or repairing the existing ones by the water shutoff, chemical stimulation, reperforating other zones, sidetracking to other areas of the reservoir, or abandoning the existing well. Production engineers focus on challenging the well production and injection targets, increasing the injection support, selecting the best artificial lift system, cleaning the production tubing form solids, and so forth. Production engineers also review and ensure maximum surface facilities operability and maintenance, including instrumentation, control, wellhead, pipelines, separators, pumps, compressors, heaters, and so forth.

Production Optimization

Optimizing production operations may—with minimum effort—involve ranking the opportunities for production increase from the production system (reservoirs, wells, surface equipment). Also, to sustain lean operations, production operations require the constant review of processes, management systems, the adaptation of people, and applications of smart technology. Often, decisions must be made despite the uncertainties of well performance, subsurface response, equipment failure rates, and downstream demands. The heterogeneity of information and the complexity of current assets implies an iterative approach to identify viable opportunities. Managing production optimization opportunities involves the management of risk and uncertainty.

Production optimization is a continuous improvement process regarding production performance and asset operations [3]. Continuous production optimization means that expected performance is frequently challenged by updating an optimal forecast with upper-level targets and current asset status (see Figure 7-4). This is achieved by applying actions that close the gap that exists between actual and expected performance. Faster surveillance loops compare actual vs. expected performance to determine minute-by-minute, hourly and daily gaps. A slower surveillance loop updates the asset's expected performance.

The main input in a continuous production optimization context includes (i) field data for calibration and validation of production models, (ii) field data for establishing asset status, (iii) wells and pipeline geometry, and (iv) prior process knowledge. The main output is (i) model response indicating expected or actual performance, (ii) model mismatch indicating field diagnosis, (iii) manipulated variables movements indicating optimal solution, and (iv) optimized forecasts and what-if scenarios.

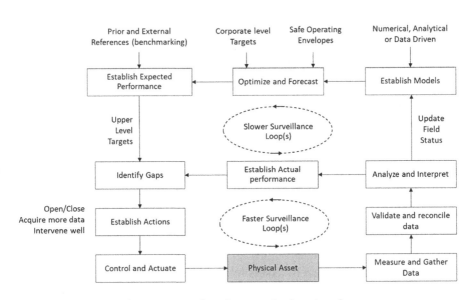

Figure 7-4. *Continuous production optimization loops*

Because of the complexity of the production system (see Figure 7-1), there are numerous decisions that an operator can make to find optimum performance. These decisions can be grouped into short-term decisions, such as adjusting the well's operating point and establishing well service requirements, and long-term decisions, such as well count, spacing and surface facility capacities.

Multivariable optimization techniques are used in many ways in the oil industry to support decisions and to accomplish daily tasks, such as resource scheduling, optimum history matching of reservoir parameters, optimal well location and spacing, production parameter settings [4] [5], and optimization of the displacement efficiency or recovery factor [6]. Such optimization uses models that are generated offline by using first principles or semi-empirically by using data acquired in the field.

Despite its extensive use in the petrochemical industry, multivariable optimization has not penetrated much of the upstream industry. To the extent used, it lacks the connection with the real field and does not consider the inclusion of dynamic data for the continuous updating of models.

Predicting Well Rate Potential

Well rate potential prediction is one key activity to assess well performance needed to establish any gap concerning current measurements. Traditionally, steady-state liquid rate (Q_{Liquid}) is predicted by using the fundamental equation (Equation 6.1) of fluid flow through porous media [7] for a vertical well, fully penetrating a single-layer, open-hole and circular drainage area (see Figure 7-5) and the pressure drop across the production system from the bottomhole to the surface.

$$Q_{Liquid} = \left(\frac{k_{ro}}{\mu_o \beta_o} + \frac{k_{rw}}{\mu_w \beta_w} \right) \left(\frac{kh}{\ln \frac{r_e}{r_w} + S} \right) \left(P_{Res} - P_{fbhp} \right). \qquad (7.1)$$

Well rate potential Q_{Liquid} depends on the rock-fluid properties including the relative permeability to oil (k_{ro}) and water (k_{rw}), the rock properties including absolute permeability (k) and thickness (h), the fluid properties (oil and water viscosities (μ_o, μ_w) and oil and water formation volume factors (β_o, β_w), the static reservoir pressure (P_{Res}), the wellbore architecture including the wellbore radius (r_w) and reservoir drainage radius (r_e) and skin (S). The reservoir inflow performance relationship (IPR), as shown in Figure 7-2, can be obtained by varying flowing bottom hole pressure (P_{fbhp}) from zero to static reservoir pressure (P_{Res}).

P_{fbhp} is a function also of Q_{Liquid}, fluid properties (i.e., initial solution gas, Rsi; oil density, °API). The pressure drop in the outflow system depends on the wellhead backpressure (P_{sep}), choke ID, internal tubing diameter (ID_{tubing}), tubing length ($TVDSS_{well}$), wellbore deviation angle, wellbore length (MD_{well}), and the fluids inside the tubing (Q_{Oil}, Q_{wat}, Q_{gas}).

Increasing Well Awareness with Virtual Sensors

On many occasions, direct physical measurements are unavailable or inconsistent with process performance due to sensor failure, technology availability, instrument reliability, or economic reasons. In such situations, virtual sensors are convenient replacements of physical sensors that use available data during well-known conditions in instrumented wells to predict other measurements [8]. Many technologies have been developed to estimate rates and pressures from other indirect measurements (e.g., virtual metering or soft sensors).

For example, wellhead pressure (WHP) and temperature (WHT) are related to flow line pressure (FLP) for a specific choke diameter; therefore, it is convenient to establish a machine learning model among these four variables so that it can serve as a replacement whenever needed. Since most of the oil and gas wells exhibit a nonstationary process, where the boundary conditions may change along with the life of the same, it is probably safe to recognize that the virtual sensor model may not be valid throughout the life of the well, but for a specific period of time.

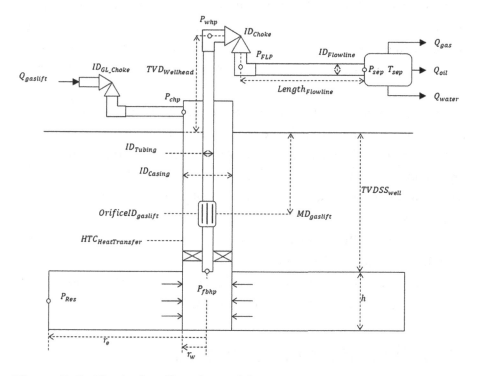

Figure 7-5. *Vertical well with gas-lift*

Virtual Rate Metering

Well surveillance is essential for reservoir characterization, managing production potential, and selecting activities to enhance production. Key to surveillance is to understand well flowing conditions (i.e., flow rates and flowing status). Typically, well rates (i.e., oil, water, and gas) are not directly measured all the time; however, with virtual sensors, it is possible to build and implement continuous *well rate estimation* (WRE), also known as *virtual rate estimators* (VRE). Virtual rate metering is one type of a virtual sensor [9].

One class of virtual rate estimators that is very well-known in the oil and gas industry is derived from physics-based nodal analysis models. This rate estimation requires consistent pressure-volume-temperature (PVT) data, fit-for-purpose production well tests, and reliable sensors [10]. In these kinds of models, missing data, biased data, or failing sensors may break the rate estimation, and a new calibration is required. In addition, sensor input uncertainty and rate estimation confidence are commonly overlooked in these approaches [11].

Another approach is to focus on data-driven VRE, enabled by machine learning algorithms. The process involves prediction models for generating instantaneous predictions of multiphase flow rates, cumulative production, and other quantities of interest, such as GOR, WCT, using real-time sensor data at the surface, and historical production, and well test data.

Data-driven VRE produces encouraging results when measured data is sufficient, including frequent well tests (e.g., 8–12 per year), permanent wellhead, and flowline sensors (pressure, temperature). The availability of downhole sensors is desirable but not required.

Predicting Well Rates from Indirect Measurements

Consider a naturally-flowing well with more than 2000 flow tests available. A machine learning model for predicting virtual rates—including oil, water, and gas—is built with half of the tests available to predict the other half. The input data refers to available well measurements, including bottomhole pressure (BHP), wellhead pressure (WHP), wellhead temperature (WHT), separator pressure (Psep), separator temperature (TSep), and chokes internal diameter (ID_Choke). The predicted output includes the oil, water, and gas rates as a function of all the given measurements.

- Input data: "Well_Rates.csv"

- Python code: "Chapter07_Well_Rate_Predictor_RF_ GBM.ipynb"

The notebook provides simplified code for the oil predictor using Random Forest (RF) regression method (i.e., rate = f(BHP, WHP, WHT, Tsep, Psep, Choke)). Additional virtual metering functions are available in the provided file.

```
df = pd.read_csv("./data/Well_Rates.csv")
X = df[["BHP", "WHP", "WHT", "Tsep", "Psep", "Choke_in"]]
y = np.array(df[["Qoil"]].values).reshape(-1, )
scaler = MinMaxScaler()  # Scale Data
```

Unsupervised classification (e.g., DBSCAN, GMIX) over the normalized PCA data can select clean data prior to feeding to the regression step.

```
X_scaled = scaler.fit_transform(X)
pca = PCA()
X_pca = pca.fit_transform(X_scaled)  # Perform PCA
pca_df = pd.DataFrame(X_pca)
# Cluster using DBSCAN
clustering = DBSCAN(eps=0.5, min_samples=12).fit(X_pca)
pca_df["DBSCAN"] = clustering.labels_ + 1
# Keep only the rows where DBSCAN=0
clust_df = pca_df[(pca_df["DBSCAN"] == 1)]
# Delete text columns "DBSCAN"
clust_df.drop(columns=["DBSCAN"],inplace=True)
# Merge data
X_merge = X.merge(pca_df,left_index=True,right_index=True)
X_merge["Labelled Y"] = y
```

Create training data and split the set into training and validation data sets.

```
# Bring only the Xrows selected as per Clust_df -->
# X.iloc[clust_df.index.tolist(),:]
# Bring only the Yrows selected as per Clust_df -->
# y[clust_df.index.tolist()
train_X, test_X, train_y, test_y = train_test_split(X.
iloc[clust_df.index.tolist(),:],y[clust_df.index.tolist()],
test_size=0.5, random_state=123)
```

Build a RF estimator.

```
# Random Forest Regression
model_RF = RF(n_estimators=200, random_state=123)
model_RF.fit(train_X, train_y)
```

Predictions for training and validation data sets.

```
pred_y_RF_test = model_RF.predict(test_X) # Blind test
pred_y_RF_train = model_RF.predict(train_X)
pred_y_RF_all = model_RF.predict(X.iloc[clust_df.index.
tolist(),:])
# Compute  r2 scores
r2_RF_train = r2_score(train_y, pred_y_RF_train)
r2_RF_test = r2_score(test_y, pred_y_RF_test)
r2_RF_all = r2_score(y[clust_df.index.tolist()], pred_y_RF_all)
```

Include calculations with the original data set so they can be plotted.

```
# RF to X_merge file
all_pred = np.array([np.nan for i in range(len(X_merge))])
all_pred[clust_df.index.tolist()] = pred_y_RF_all
X_merge["Predicted_RF"] = all_pred
X_merge.to_csv("X_merge.csv", index=False)
```

Prepare plots.

```
# Xplot of predicted and actual values
fig = plt.figure(figsize=(6,6))
ax = plt.axes()
plt.scatter(y[clust_df.index.tolist()], pred_y_RF_all, c="blue")
plt.scatter(test_y, pred_y_RF_test, c="red")
ax.set(xlabel="Test Data", ylabel="RF Model Prediction",
title="Oil Rate RF Predictor Model")
plt.show()
```

The same procedure can be repeated with the GBM estimator or any other algorithm that you prefer. The results of RF and GBM regression methods are shown in Figure 7-6.

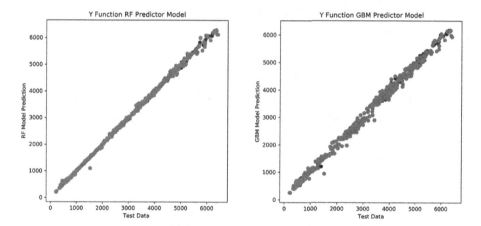

Figure 7-6. *Well virtual rate estimator for RF and GBM methods (red is training data, blue is blind test data)*

The regressor scores are as follows.

- R^2 for train data using random forest is 0.999838.

- R^2 for blind test data using random forest is 0.998584.

- R^2 for all data using random forest is 0.999230.

235

- R^2 for train data using GradientBoostingRegressor is 0.995428.

- R^2 for blind test data using GradientBoostingRegressor is 0.992874.

- R^2 for all data using GradientBoostingRegressor is 0.994193.

Notice that for there are six available measurements ([BHP, WHP, WHT, Tsep, Psep, Choke_in) the engineer can implement for multiple virtual rate calculators, also known as control volume, just by changing the input features in training matrix X and output vector Y.

One way is to "uncomment" the required X and Y combination .

```
# Create input feature and target
#X = df[["WHP", "Psep"]] # flow line performance
#X = df[["WHP", "Psep", "Choke_in"]] # flow line and choke
performance
#X = df[["BHP", "WHP"]] # tubing vertical lift performance (VLP)
#X = df[["BHP", "WHP", "Choke_in"]] # VLP and choke performance
#X = df[["WHP", "WHT", "Tsep", "Psep"]] # flow line P&T
performance
#X = df[["WHP", "WHT", "Tsep", "Psep", "Choke_in"]] # choke
performance                                      '
#y = np.array(df[["Qoil"]].values).reshape(-1, )
#y = np.array(df[["Qwater"]].values).reshape(-1, )
#y = np.array(df[["Qgas"]].values).reshape(-1, )
```

Another convenient method is to build independent virtual metering functions and run all of them simultaneously (i.e., Y1oil = f(X1), Y1water = f(X1), Y1gas = f(X1), Y2oil = f(X2), Y2water = f(X2), Y2gas = f(X2), ..., etc.).

```
# Create input feature and target
X0 = df[["BHP", "WHP", "WHT", "Tsep", "Psep", "Choke_in"]]
X1 = df[["WHP", "Psep"]]
X2 = df[["WHP", "Psep", "Choke_in"]]
X3 = df[["BHP", "WHP"]]
X4 = df[["BHP", "WHP", "Choke_in"]]
X5 = df[["WHP", "WHT", "Tsep", "Psep"]]
X6 = df[["WHP", "WHT", "Tsep", "Psep", "Choke_in"]]
Y1 = np.array(df[["Qgas"]].values).reshape(-1, )
Y2 = np.array(df[["Qoil"]].values).reshape(-1, )
Y3 = np.array(df[["Qwater"]].values).reshape(-1, )
```

The advantage of this approach is that all 7 input matrices (X0 to X6) and for each rate (Y1, Y2, Y3) have been built with different physical process assumptions; for example flow across the flow line, flow across the choke or flow across the tubing.

The obtained models tend to agree when the process conditions are stable, and the models are well-calibrated. But they are also powerful tools for identifying exceptions in the system; these models diverge when process conditions are different from the original assumptions; for example, increasing water cut or increasing GOR.

Predicting Well Rates for Gas-lift Wells

A gas-lift well (see Figure 7-7), whose properties are shown in Table 7-1, has about 12,538 flow test data points. A machine learning model for predicting well rates (including oil, water, and gas) will be built. The input data refers to available measurements at the well, including bottomhole pressure (BHP), wellhead pressure (WHP), wellhead temperature (WHT), flowline pressure and temperatures (FLP, FLT), and choke internal diameter (ID_Choke). The predicted output includes the oil, water, and gas rates as a function of all the given measurements.

Table 7-1. *Vertical Well with Gas-Lift Properties*

	Average	**Standard deviation**
Oil Gravity, °API	36.3	0.2
Oil Initial Solution Gas, Rsi, SCF/bbl	516.7	3.0
Gas Specific Gravity,	0.8	0.0
Water Salinity, ppm	230,010	1,330.2
Reservoir Temperature, °F	240	1.4
Reservoir Pressure, psi	4,347	376.3
Tubing Length, ft	12,269	7.1
Casing ID, inches	6.2	0.0
Tubing ID, inches	3.1	0.0
Gas lift Orifice Depth, ft	12,069	7.1
Flowline ID, inches	3.1	0.0
Flowline Length, ft	10	0.1
Temperature Surface, °F	80	0.5
Heat Transfer Coefficient, BTU/(hr•ft²•°F)	9.2	1.6

- Input data: "Well_Choke_GL.csv"

- Python code: "Chapte07_GL_Well_Rate_and_Pressure_MLP_Predictor.ipynb"

Python code for the oil, water, and gas rate prediction is implemented using the multilayer perceptron (MLP) regression method (i.e., rate = f(BHP, WHP, WHT, FLP, FLT, ID_Choke, ID_orifice, Q_Gas_lift)).

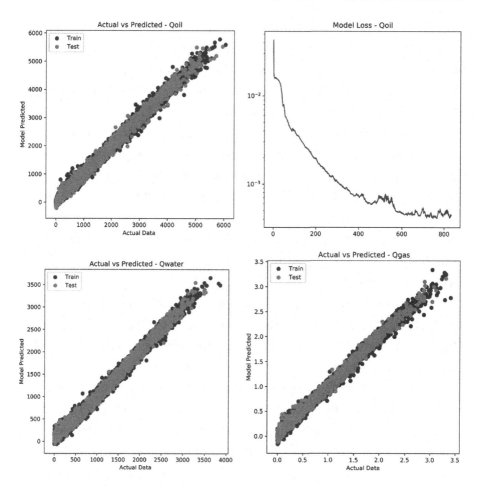

Figure 7-7. *Well rate (oil, water, and gas) estimator using MLP algorithm for gas-lift well*

Predicting Well Failures

One of the production engineer's tasks is to establish practices and methods to proactively detect and prevent well failures, to recommend actions on the well (and facilities) operating envelopes. Reservoir, near wellbore environment, downhole equipment, and any component of the surface production system may fail at any point in time, leading to considerable unplanned downtime, production losses, and uncontrolled operation expenditures. The proactive identification and prevention of well failures lead to appropriate maintenance scheduling which ultimately reduces inappropriate repairs, minimizes downtime, and subsequently improves operation efficiency [12].

Engineers face challenges in identifying candidates and ranking opportunities for production enhancement purely from performance data. Standard production data management and decision-making platforms poorly predict a decline in uptime, availability, and optimum operating envelope of wells. The challenges typically relate to proper system understanding, well condition diagnosis, proactive opportunity identification, efficient opportunity ranking of opportunities, and post-job performance analysis.

Systematic pattern recognition workflow to detect the well failures in sucker rod pumped (SRP) wells have been successfully implemented [12]. This led to tremendous savings in engineering tied to detecting SRP failures, as well as minimizing the production losses.

Recommending the optimum treatment method is also a necessary task to manage operating expenditures, not only by minimizing rig and equipment cost but also by bringing the well faster to production and avoiding future failure for the same cause [13]. The idea is to combine historic success and failure data, with field personnel comments, and basic business rules into a data-driven model able to recommend treatment methods that optimize design, field plan, equipment use, and job performance.

The steps to build any advisory solution that predicts well failure, include data screening, data feature engineering, root-cause-analysis, recommending actions, and feedback.

Data Screening

Data screening identifies symptoms that show where asset performance is not as good as expected. Patterns among wells are detected to identify similar behavior and reduce the complexity of the screening problem from several hundred sensors to a few categories of similar measurement types.

Data Feature Engineering

Data feature engineering may involve the generation of data features using domain-knowledge, feature extraction, and pattern recognition. The objective of this step is to allow the algorithm to act on trends rather than on actual values. Rather than acknowledging the fact that the *tubing head pressure* is 1200 kPa, for example, the advisory system should consider whether the tubing head pressure is increasing, stable, or decreasing. The rate of change in any of the input variables is only one of the features in a data stream that needs to be extracted.

Domain expert interviews and the elaboration of the surveillance logic reveal that it is a combination of features in the data streams that are typically used by experts to detect events. In only a few cases, it is an absolute value or threshold that alarms the experts; usually, it is a series of features in one signal or many signals combined.

Root Cause Analysis (RCA)

Technical analysis methods and diagnoses are applied to identify the causes of why the performance may be above or below expectations. The objective is to identify the constraint, such as whether the liquid production is limited by reservoir deliverability, by well production potential, or by facility processing limits.

Due to the ambiguity of some of the symptoms as identified in the previous step (feature engineering), the outcome of RCA is typically probabilistic, indicating the most likely causes but also possible alternative causes.

One known method of a root cause identification is Bayesian Network. An example for determining the most likely cause for declining downhole pump performance is determined by the Bayesian Network model which considers the observations and real-time measurements. The observations can be independent (e.g., pumpage and reservoir pressure) but can also be linked by a causal relationship (e.g., wellhead pressure and production rates).

In this example, the information about the pump, reservoir, fluid, measuring dropping liquid rates, and wellhead pressures, are used to determine the most likely cause which is gas ingestion; however, this does not entirely exclude mechanical problems and possible excessive pump wear (e.g., due to sand in the pump).

Recommending Actions to Improve Performance

A high-level well "health score" can be calculated from features identified in well sensors and other measurement data. Additional scores can be calculated to quantify the problem severity (i.e., problem score), the size of the opportunity if the well is fixed or improved (i.e., potential score), and so on. All scores that are calculated for a well are summed up to yield a final score. The final score can be considered a health score; the smaller the value, the less likely it is that the well shows a severe issue. The wells can then be ranked by the final score to give a list of priorities for the daily operations.

The calculated events, conditions, and recommended actions are availed for each element in the production system (e.g., for each well, for each facility). The result of how likely an event occurs multiplied by how severe the impact of an event may be (safety issue, lost production, lost time) is a score for a certain event.

Based on a definition of utility (e.g., maximize production, minimize losses, increase net present value, reduce lost time, etc.), decisions are suggested to solve the problems as identified in previous steps. The suggestion may be probabilistic. The overall utility value is a combination of various sub-utilities; therefore, a unified evaluating system must be established for every organization or even for every project, which allows a comparison of monetary and nonmonetary targets.

Predicting Poor Well Performance

One of the challenges encountered in predicting poor well performance stems from the fact that there are very few indications of a rapid decline in production during the early stages of production. However, by using unsupervised machine learning techniques, such as clustering algorithms, it could be possible to identify the wells, which may exhibit diminished oil production over a certain number of years. A methodology for a clustering-based approach is shown in Figure 7-8.

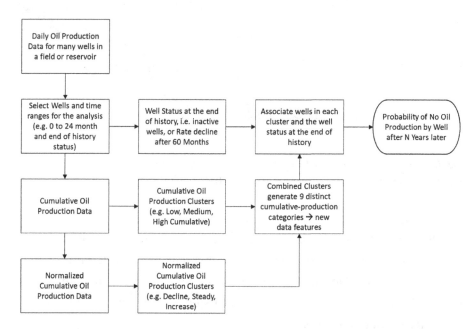

Figure 7-8. *A clustering-based methodology for identifying wells with a high probability of diminished production in future*

The daily production data from many wells during the early stages of production (e.g., the initial 24 months) is considered for computing cumulative oil production from the wells. Additionally, cumulative oil production from each well is normalized to 1 by using the cumulative production from the well at the end of the duration being considered. These two cumulative production plots can then create well clusters separately. The clustering based on cumulative oil production, shown in Figure 7-9 (left), leads to three distinct categories of wells.

- Low cumulative production wells (blue lines)

- Medium cumulative production wells (red lines)

- High cumulative production wells (green lines)

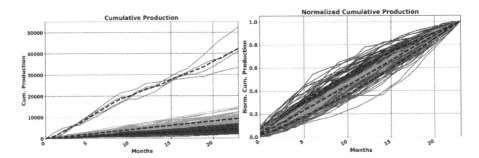

Figure 7-9. *Clustering of cumulative oil production (left) and normalized cumulative production exhibits distinct well clusters (right)*

Also, clustering based on normalized cumulative oil production shows three categories of wells (see Figure 7-9 right). The black dotted lines show the cluster mean values for each instance of time. Observing the cluster mean, it becomes clear that these three clusters denote following three different production trends:

- Declining production rate (blue lines), with decreasing slope of the cluster mean

- Steady production rate (red lines), with an almost constant slope of the cluster mean

- Increasing production rate (green lines), with an increasing slope of the cluster mean

By combining the clustering based on the cumulative oil production and normalized cumulative oil production, nine distinct well categories are identified, each belonging to a combination of cumulative oil production and production rate category (e.g., high cumulative production—increasing production rate, low cumulative production—declining production rate, etc.).

Furthermore, the number of wells from the production data analyzed demonstrated a critical decline in oil production over five years or longer time horizon. For each of the nine categories, the probability of a critical decline in oil production was calculated by considering the total number of wells in that category and the number of wells showing a critical decline in the future. Table 7-2 summarizes the computed probability that a well will be dead within the next five years.

It can be readily observed that the wells in the "high cumulative production" cluster did not show any probability of becoming inactive within the time horizon considered in this study, irrespective of the production rate cluster they belonged to.

Table 7-2. *Probability of Dead Well in Five years*

Production Rate → Cumulative Production ↓	Declining Rate	Steady Rate	Increasing Rate
Low	0.258	**0.308**	**0.357**
Medium	0.12	0.14	0.19
High	0.000	0.000	0.000

Average Water Cut and Dead Well Probability Correlation Coefficient: 0.719563

Additionally, wells in the categories with low cumulative production along with steady or increasing production rates demonstrated significantly higher probabilities of becoming inactive within the time horizon considered. Based on these observations, it is suggested that for the studied oil field, the new wells with low cumulative oil production along with steady or declining production rates should be monitored carefully. The wells falling in these two categories based on the first 24 months production data can be prioritized for the more frequent surveillance activities and production enhancement workover activities.

- Input data: "Well_Monthly_Production.csv"

- Output: "Probability of a well becoming inactive"

- Python code: "Chapter07_Well_Failure_Clustering_using_KMeans.ipynb"

- Method: k-means clustering

Managing Rate Dependent Water Cut

On many occasions, water cut (WCT) and gas/oil ratios (GOR) are rate-dependent phenomena that can be minimized at reduced rates. These phenomena can be the results of a higher mobility phase viscous fingering and/or coning.

Water coning occurs when oil-water contact surface under the well forms a cone-shaped profile due to well's pressure drawdown, overcoming oil/water gravity segregation. After a water breakthrough, coning increases water cut and reduces oil production and recovery significantly [14]. An operator may produce the well at a rate resulting in no water or gas breakthrough. This maximum production at a steady-state condition is known as the *critical oil rate*. Critical oil-rate guidelines restrict the short-term benefits; therefore, it is important to calculate it wisely, particularly in naturally fractured reservoirs (NFRs).

Since NFRs with bottom-water drive are notorious for instant water-breakthrough and extremely high water cut, producing oil well below critical oil production rate may become non-economical when the rate is too low. So, there is a need to study the feasibility of critical oil rate production in NFRs [14].

Predicting Critical Oil Rate

A critical oil rate predictor was developed by modeling a wide variety of NFRs with bottom-water using a dual-porosity-dual-permeability (DPDP) model implemented on a commercial simulator [15]. Simulated input data showed that the critical oil rate for a few NFRs could be as high as 200 bopd, which is economical.

The following steps were carried out in the development of a semi-analytical model of critical oil rate [14].

1. Simplified the NFR by representing the fracture-network as an equivalent continuous porous media.

2. Verified Chaperon's mechanistic model [16] of critical rate with the simulated critical-rate values of NFR after replacing the permeability of matrix with an equivalent permeability of NFR.

3. Calibrated the resulting mechanistic model statistically using a designed series of simulated experiments representing a wide variety of NFRs, to include the ignored physical effects.

These steps resulted in a statistically calibrated mechanistic model [16] for critical oil rate in NFRs ($q_{cr,fr}$), which can be written as Equation 7.2.

$$q_{cr,fr} = 0.0783 \times 10^{-4} \frac{k_{f,m} h_o^2}{B_o \mu_o} (\Delta \rho) \left[1 - \left(\frac{h_{op}}{h_o} \right)^2 \right] f \left(\frac{r_e}{h_o} \sqrt{\frac{k_{fv}}{k_{fh}}} \right). \qquad (7.2)$$

Where, $k_{f,m} = k_{fh} + k_{mh}$

Where, k_{fh} is the effective fracture permeability, md; k_{mh} is the matrix permeability, md; $k_{f,m}$ is the average permeability of the NFR, md; h_o is the oil-zone thickness, ft; r_e is the reservoir radius, ft; μ_o is the oil viscosity, cp; B_o is the oil formation volume factor; h_{op} is the perforated interval, ft; $\frac{k_{fv}}{k_{fh}}$ is the anisotropy ratio of the fracture network; $q_{cr,fr}$ is the critical oil rate of on-fracture well in NFR.

For determining the empirical component of the model (Equation 7.1), $f \left(\frac{r_e}{h_o} \sqrt{\frac{k_{fv}}{k_{fh}}} \right)$, we design the series of simulated experiments by varying well-reservoir properties, including effective fracture permeability, matrix permeability, mobility ratio, reservoir radius, anisotropy ratio, and penetration ratio [17]. The Box-Behnken experimental design [18] [19], a three-level factorial design, creates the matrix of simulated experiments, which capture nonlinear effects while minimizing the number of experiments. To set-up the regression model, Equation 7-2 is rewritten in the form of a normalized critical rate ($\widehat{q_{cr,fr}}$), as in (Equation 7.3).

$$\widehat{q_{cr,fr}} = \frac{q_{cr,fr}}{0.0783 \times 10^{-4} \frac{k_{f,m} h_o^2}{B_o \mu_o} (\Delta \rho) \left[1 - \left(\frac{h_{op}}{h_o} \right)^2 \right]} = f \left(\frac{r_e}{h_o} \sqrt{\frac{k_{fv}}{k_{fh}}} \right). \qquad (7.3)$$

After running the linear regression, the normalized critical rate is correlated as a function of aspect ratio, $\frac{r_e}{h_o} \sqrt{\frac{k_{fv}}{k_{fh}}}$, whose regression coefficients are shown in Table 7-3.

Table 7-3. *Regression Model of Empirical Parameter,* $f\left(\dfrac{r_e}{h_o}\sqrt{\dfrac{k_{fv}}{k_{fh}}}\right)$

Variable	Coefficients	t Value	Pr > ltl
Intercept	0.728	214.7	0
$\dfrac{1}{\dfrac{r_e}{h_o}\sqrt{\dfrac{k_{fv}}{k_{fh}}}}$	1.997	18.4	0

The regression model demonstrates a good fit with experimental data at an R-squared value of 0.85. Hence, the final equation of the critical rate for an on-fracture well in NFR can be written as Equation 7.4.

$$q_{cr,fr}=0.0783\times10^{-4}\frac{k_{f,m}h_o^2}{B_o\mu_o}(\Delta\rho)\left(1-\left(\frac{h_{op}}{h_o}\right)^2\right)\left(0.728+\frac{1.997}{\dfrac{r_e}{h_o}\sqrt{\dfrac{k_{fv}}{k_{fh}}}}\right). \quad (7.4)$$

The implementation of the critical oil rate as a function of the input parameters.

- Input data: "Critical_Oil_Rate.csv"

- Output: "Critical Oil Rates"

- Python code: "Chapter07_Critical_Oil_Rate_Predictor_using_OLS.ipynb"

- Method: Multivariable linear regression methods

```
import statsmodels.api as sm
from statsmodels.tools.tools import add_constant
import pandas as pd
import statsmodels
df = pd.read_csv("Critical_Oil_Rate.csv")

Y = df.normalized_criticalrate.values
x = df.Empirical_component.values
X = sm.add_constant(x) # adding a constant

model = sm.OLS(Y, X).fit()
predictions = model.predict(X)
model.summary()
```

OLS Regression Results

Dep. Variable:	y	R-squared:	0.850
Model:	OLS	Adj. R-squared:	0.847
Method:	Least Squares	F-statistic:	339.5
Date:	Wed, 15 Jul 2020	Prob (F-statistic):	2.21e-26
Time:	21:36:44	Log-Likelihood:	170.01
No. Observations:	62	AIC:	-336.0
Df Residuals:	60	BIC:	-331.8
Df Model:	1		
Covariance Type:	nonrobust		

	coef	std err	t	P>|t|
	[0.025	0.975]		
const	0.7277	0.003	214.698	0.000
	0.721	0.734		
x1	1.9969	0.108	18.426	0.000
	1.780	2.214		

Omnibus:	119.548	Durbin-Watson:	1.967
Prob(Omnibus):	0.000	Jarque-Bera (JB):	5122.268
Skew:	-6.193	Prob(JB):	0.00
Kurtosis:	45.772	Cond. No.	53.9

Recommending Optimum Well Spacing in NFRs

Ultimate water cut defines the theoretical maximum water cut in a reservoir with water coning, which determines the maximum economically producible oil in a bottom-water reservoir. During the production of such reservoirs, there is an initial rapid increase in water cut during the water cone buildup stage, which is later followed by the stabilization of water cut (ultimate water cut) in an uncontained oil-zone (infinite oil-zone). However, in a multi-well reservoir where each well is bounded by a nonflow boundary (NFB), water cut would never stabilize but is replaced by a slow water cut increase/advancement rate. This is because of the depletion of oil in a limited boundary reservoir resulting in a slow increase of water cut.

If the well's drainage area is large enough that the size exceeds the lateral length of the water cone (at the end of water cone buildup (expansion) stage), the late-time, water cut, slow-advancement stage for the NFB system is controlled mostly by the oil-pay depletion, as the effect of the water cone expansion becomes negligible and can be ignored.

Late-time water cut may render the oil production non-economical, so the well's recovery is mostly controlled by the time when the cone expansion ends. This implies that the optimum well's drainage radius should be equal to the lateral length of the water cone, which enables the well's maximum oil recovery. This concept is demonstrated in Figure 7-10. This minimum inter-well spacing is defined as threshold well spacing, or two times the threshold drainage radius of a well [14].

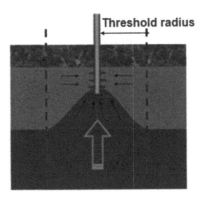

Figure 7-10. *Threshold drainage radius equal to the water cone lateral length (threshold radius) enabling maximum well recovery*

Predicting Threshold Reservoir Radius

Threshold radius can be determined statistically using a large number of simulated experiments for a variety of well/reservoir system, by varying the properties including horizontal permeability (k_h), anisotropy ratio ($\frac{k_v}{k_h}$) and penetration ratio ($\frac{h_{op}}{h_o}$), oil-pay thickness (h_o), mobility (M) ratio and aquifer thickness (h_w). The Box and Behnken method can design a matrix of simulated experiments [18]. The design stipulates a total of 54 different simulation runs (reservoir/aquifer systems) to determine the threshold radius by systematically increasing the size of the drainage area until it becomes equal to the lateral length of maximum water cone. The results are then used to develop an empirical formula for threshold drainage radius, r_{eTh} using the multiple second-order polynomial regression.

The resulting formula for predicting threshold reservoir radius is

$$r_{eTh} = 14920.6 - 3563.7M - 68.5h_w - 34586\frac{k_v}{k_h} + 99.65h_o + 288M^2$$

$$+ 0.107h_w^2 + 31290\left(\frac{k_v}{k_h}\right)^2 + 1.5Mh_w - 0.55Mk_h - 10.13Mh_o$$

$$- 0.087h_wh_o + 0.1k_hh_o.$$

(7.5)

The following is the implementation of the threshold reservoir radius predictor as a function of the input parameters.

- Input data: Threshold_reservoir_radius.csv

- Output: "Threshold Radius values"

- Python code: "Chapter07_Threshold_Radius_Predictor_using_OLS.ipynb"

- Method: Multivariable linear regression methods

```python
import statsmodels.api as sm
from statsmodels.tools.tools import add_constant
import pandas as pd
import statsmodels
from statsmodels.stats.outliers_influence import variance_
inflation_factor
df = pd.read_csv("Threshold_reservoir_radius.csv")
Y = df.threshold_radius.values
x = df.drop(['threshold_radius' ], axis=1)
vif = pd.DataFrame()
vif = [variance_inflation_factor(x.values, i) for i in range(x.
shape[1])]
print(vif)
X = sm.add_constant(x) # adding a constant
model = sm.OLS(Y, X).fit()
predictions = model.predict(X)
model.summary()
```

OLS Regression Results

Dep. Variable:	y	R-squared:		0.693
Model:	OLS	Adj. R-squared:		0.654
Method:	Least Squares	F-statistic:		7.67
Date:	Sat, 02 May 2020	Prob (F-statistic):		1.39e-10
Time:	23:17:49	Log-Likelihood:		-512.69
No. Observations:	54	AIC:		1039.
Df Residuals:	47	BIC:		1053.
Df Model:	6			
Covariance Type:	nonrobust			

	coef	std err	t	P>|t|	[0.025	0.975]
const	6094.0420	1815.138	3.357	0.002	2442.455	9745.629
Oil-zone thickness	71.2688	10.830	6.581	0.000	49.482	93.055
Mobility	-888.1647	144.831	-6.132	0.000	-1179.528	-596.802
Aquifer thickness	-10.5367	2.554	-4.125	0.000	-15.675	-5.398

Horizontal permeability	5.0949	2.719	1.874	0.067	-0.375	10.565
Penetration ratio	-275.2788	2302.091	-0.120	0.905	-4906.489	4355.932
Anisotropy ratio	-1728.2702	1189.820	-1.453	0.153	-4121.879	665.339

===

Omnibus:	7.561	Durbin-Watson:	1.358
Prob(Omnibus):	0.023	Jarque-Bera (JB):	6.630
Skew:	0.762	Prob(JB):	0.0363
Kurtosis:	3.791	Cond. No.	1.73e+03

===

255

Summary

In this chapter, we provided a brief overview of production engineering. We also discussed production optimization and the formulation for predicting well rate performance. Using realistic synthetic data generated with first-principles models, we presented you with the machine learning approach to virtual metering. The virtual metering example demonstrates how sensor data obtained from the oil wells can help us continuously monitor production from a well. A section of this chapter also provided example implementation of machine learning algorithms for predicting impending well failure, using the historical production data.

In the last sections of the chapter, we discussed the problem of predicting critical oil rate and the determination of optimal well spacing in the naturally fractured reservoirs. We hope that you were able to develop a good understanding of various ways to use sensor data and production data in conjunction with the machine learning algorithms to solve production engineering problems, which are observed during the production operations.

Acknowledgments

We would like to thank Samir Prasun, PhD, for his valuable suggestions and contributions to this chapter.

References

[1] A. Shields, S. Tihonova, R. Stott, L. Saputelli, Z. Haris, and A. Verde, "Integrated Production Modelling for CSG Production Forecasting," SPE Asia Pacific Unconventional Resources Conference and Exhibition, Brisbane, Australia, 2015.

[2] L. Saputelli, C. Bravo, M. Nikolaou, C. Lopez, R. Cramer, S. Mochizuki, and G. Moricca, "Best Practices and Lessons Learned After 10 Years of Digital Oilfield (DOF) Implementations," SPE Kuwait Oil and Gas Show and Conference, Kuwait City, Kuwait, 2013.

[3] W. Omole, L. Saputelli, J. Lissanon, O. Nnaji, F. Gonzalez, G. Wachel, K. Boles, E. Leon, N. Parekh, N. Nguema, J. Borges, and P. Hadjipieris, "Real-time Production Optimization in the Okume Complex Field, Offshore Equatorial Guinea," SPE Digital Energy Conference and Exhibition, The Woodlands, TX, 2011.

[4] W. Ramirez, "Optimal Injection Policies for Enhanced Oil Recovery: Part 1 Theory and Computational Strategies," *Society of Petroleum Engineers Journal*, June 1984.

[5] W. Liu and W. Ramirez, "Interactive Personal Computer Optimal Control Calculations For Steamflooding," *Society of Petroleum Engineers Journal*, January 1992.

[6] B. Sudaryanto and Y. Yortsos, "Optimization of Displacements in Porous Media Using Rate Control," SPE Annual Technical Conference and Exhibition, New Orleans, LA, 2001.

[7] H. Darcy, *Les fontaines publiques de la ville de Dijon*, Paris: Dalmont, 1856.

[8] H. Melbø, S. Morud, B. Bringedal, R. van der Geest and K. Stenersen, "Software that Enables Flow Metering of Well Rates with Long Tiebacks and With Limited or Inaccurate Instrumentation," Offshore Technology Conference, Houston, TX, 2003.

[9] American Petroleum Institute, *API Recommended Practice 86: Recommended Practice for Measurement of Multiphase Flow*, Washington, DC, 2005.

[10] A. Petukov, L. Saputelli, J. Hermann, A. Traxler, K. Boles, O. Nnaji, B. Vrielynck, and D. Vegunopal, "Virtual Metering System Application in the Ceiba Field, Offshore Equatorial Guinea," SPE Digital Energy Conference and Exhibition, The Woodlands, TX, 2011.

[11] H. Melbø, B. Bringedal, N. Hall, S. Morud, E. Birkemoe, and C. Smith, "Uncertainty Based Allocation Using Virtual Multiphase Flow Metering," 22nd North Sea Flow Measurement, St. Andrews, Scotland, 2004.

[12] F. Liu and A. Patel, "Well Failure Detection for Rod Pump Artificial Lift System Through Pattern Recognition," International Petroleum Technology Conference, Beijing, China, 2013.

[13] C. S. Popa, "Artificial Intelligence for Heavy Oil Assets: The Evolution of Solutions and Organizational Capability," SPE Annual Technical Conference and Exhibition, San Antonio, TX, 2012.

[14] S. Prasun and A. Wojtanowicz, "Determination and Implication of Ultimate Water cut in Well-Spacing Design for Developed Reservoirs with Water Coning," *Journal of Energy Resources Technology,* 2018.

[15] CMG, "IMEX," 2020. [Online]. Available: `https://www.cmgl.ca/imex`.

[16] I. Chaperon, "Theoretical Study of Coning Toward Horizontal and Vertical Wells in Anisotropic Formations: Subcritical and Critical Rates," SPE 61st ATCE, New Orleans, 1986.

[17] S. Prasun and A. Wojtanowicz, "Semi-Analytical Prediction of Critical Oil Rate in Naturally Fractured Reservoirs with Water Coning," *Journal of Petroleum Science and Engineering,* pp. 180, 779–792, 2019.

[18] M. Cavazzuti, *Optimization Methods: From Theory to Design Scientific and Technological Aspects in Mechanics*, Berlin/Heidelberg: Springer-Verlag, 2013.

[19] S. Ferreira, "Box-Behnken Design: An Alternative for the Optimization of Analytical Methods," *Analytica Chimica Acta,* p. 597(02), 2007.

[20] R. Roberts and R. Flin, "Unlocking the Potential: Understanding the Psychological Factors That Influence Technology Adoption in the Upstream Oil and Gas Industry," *SPE Journal,* vol. 25, no. 1, pp. 515–528, February 1, 2020.

[21] A. Phillips and B. Briegendal, "Application of Virtual Flow Metering as a Backup or Alternative to Multiphase Flow Measuring Devices," Society of Underwater Technology, 2006.

CHAPTER 8

Opportunities, Challenges, and Future Trends

During the next three decades (2020 to 2050), gross domestic product (GDP) per capita is expected to increase an average of 2% to 4% per year. Because of the shifting of many manufacturing sites to Africa, South Asia, and India, non-OECD countries will experience two to four times more growth than OECD countries [1]. The oil and gas industry will continue to exhibit an era of growth. More than half of the global energy demand in 2018 was supplied by the oil and gas industry [1], and it will continue to provide up to 50% of the energy mix through 2050. These growth estimates were generated prior to COVID-19 pandemic effects of 2020.

However, three interconnected challenges are disturbing the oil and gas industry: profitability, environmental impact, and alternative energies; these challenges are imposing constraints for reserves access, production growth, and sustainability.

- As the ability to build capacity is greater than the demand increase, commodity prices and profitability tend to be limited. During periods of higher oil and gas prices (i.e., above 50 USD/bbl), more countries accelerate their potential to bring high-cost resources. Low-cost producing provinces can survive regardless of global price trends. To find, develop, and produce at lower unit costs, operating companies are expecting to get more throughput with fewer resources, both human and technical.

- The oil and gas industry is perceived as the "black sheep" of all the energy industries. The sustainability of hydrocarbon energy is jeopardized unless the industry continuously invests in the reduction of its environmental footprint, both from leaner surface operation sites and lower-pollutant processed fuels. The reuse of CO_2 for enhanced oil recovery offers an option for sustainability.

- Hydrocarbons are mainly used in industrial sectors, transportation, and power generation. As the end consumption shift to electricity use, several alternate energies (i.e., solar, nuclear, and wind) are becoming more popular to fill in this demand with incentivized access for new consumers, continuously lowering unit costs and cleaner options. Additionally, environmental protection policies in OECD and non-OECD countries favor solar, nuclear, and hydraulic energies. This situation reduces the rate of increase in the hydrocarbon mix of global demand.

During the next decades, the oil and gas industry will redefine itself to consolidate its position in the long term only through innovation. The transformation of the traditional hydrocarbon industry with Oil and Gas 4.0 can be the game-changer for persisting over the next 30 to 50 years; digitalization can provide a unique foundation for innovation while enhancing general public awareness and human life quality. Automated and robotized field operations and office processes can enable cost reductions while improving people's life and society's quality.

The objective of this chapter is to outline the opportunities and challenges the industry has in the application of machine learning and artificial intelligence. It also outlines the expected future trends in various use cases and technologies that could shape the way the oil and gas industry navigates over the next three decades.

Challenges and Opportunities for Applying Machine Learning

Many challenges arise for the oil and gas industry to be bearable in the following decades. These can be listed from various perspectives: shareholders' perspective (profitability and business sustainability), people's happiness and efficiency, business process, and technology (see Figure 8-1).

From the shareholder's perspective, profitability and sustainability are the most prevalent goals. They are ultimately influenced by price, product demand, net unit costs, safety performance, employee satisfaction, and environmental footprint. From the people's perspective, the common challenges refer to how human goals are achieved (e.g., compensation, development, and growth) and how people contribute to the company's goals (e.g., information sharing, innovation, continuous improvement mindset, efficiency). From the business process point of view, companies deploy efficient methods to continuously streamline product transformation from finding, developing, producing, and transforming hydrocarbons.

Process Challenges

- Stress to increase profitability, reducing cost from less human resources.
- Managing downtime.
- Disconnected business processes.
- Disconnected functions.
- Potential to increase revenue is constrained by market size.

People Challenges

- Low availability and poor adequacy of the new workforce
- Separation among subsurface and digital disciplines.
- Silo mentality across multi-disciplinary teams and organizations
- Managing unplanned and ad-hoc

Technology Challenges

- Requirement to continuously monitor and optimize asset performance.
- Demonstrating commitment to environmental concerns
  Deliver a safer workplace with less human-intensive actions

Figure 8-1. Oil and gas industry challenges and opportunities for applying machine learning and artificial intelligence

Process Challenges and Mitigations

The following highlights the most relevant process-related challenges and suggested mitigating options.

- **Challenge:** Companies are continuously stressed with increasing profitability, mostly by reducing cost by employing fewer human resources, while leading to longer than usual working periods and less safe working sites. Future generations will have to do more with less, hence yielding more results per unit time and less unit cost.

Mitigation Options

- Automate drilling site operations using robotized operations and leveraging advanced data analytics for event detection focusing on abnormal critical events (i.e., stuck pipe and lost circulation).

- Automate production monitoring adding continuous data analytics of wellsite parameters while enabling routine well analysis via exception-based surveillance (EBS) focusing on critical well status.

- Implement well and reservoir performance analytics for proactive opportunity identification while focusing on high-value-added activities [2].

- AI-assisted complex decision-making with an automated routine-action recommendation.

- **Challenge**: Managing downtime. One key component of the oil and gas production capacity are the time the facility is to be shut down and the instances when it is not able to deliver its intended potential. A key challenge is to maintain these periods as low-and-predictable as possible. Unintended, unplanned shutdown time may derail the unit operating costs significantly.

Mitigation Options

- Implement data analytics and business processes to predict unplanned events, track downtime, production losses (a.k.a. *production deferral*), and causes of planned vs. unplanned downtime.

- Implement event classification to enable root-cause-analysis (RCA) of all unplanned downtime events.

- **Challenge**: Disconnected business processes, and difficulties in a functional organizational design due to the challenges originating from a silo mentality. For some integrated oil and gas companies, key business processes are typically disconnected across functions and within the functions.

Mitigation Options

- Implement KPIs across functions and share them across the multiple levels of the organization using business analytics dashboards.

- Define key roles and responsibilities that are shared across functions, and share them.

- **Challenge**: At the upstream level, production operations, reservoir management, field development, and drilling are typically disconnected functions. This hinders global optimization of the opportunities for enhancing business performance. Is it better to repair a high water producing well or drill a new drain?

Mitigation Options

- Implement an integrated well performance enhancement opportunity identification and ranking process that involves multi-objective constrained global optimization objective functions looking at short- and long-term goals.

- Faster and reliable forecasts based on integrated right-physics approaches leveraging deep learning.

- **Challenge**: At the corporate level, upstream and downstream have very different KPIs. Although the sale of raw oil is more profitable, the potential to increase revenue is typically constrained by market size; therefore, the option to increase profit is to maximize the revenues from downstream products without jeopardizing the upstream obligations.

Mitigation Options

- Implement a corporate level planning and scheduling process that involves multi-objective constrained global optimization looking at short- and long-term goals. This involves using gradient-free methods with advanced evolutionary optimization techniques.

People Challenges and Mitigations

The following highlights the most relevant people-related challenges and suggested mitigating options.

- **Challenge**: Low availability and poor adequacy of the new workforce intensify sustainability goals. Young professionals may not have the required level of competencies to address the impending challenges and require long times to get trained.

Mitigation Options

- Incorporate hybrid academic syllabuses in undergraduate and graduate-level programs.

- Offer public data sets allowing students and research community to familiarize themselves with typical data sets, setting industry expectations for standard analytics and machine learning problems.

- Implement *enterprise document management* (EDM) and knowledge search leveraging *natural language processing* (NLP) to optimize resource utilization through better technical information sharing and knowledge leverage.

- Implement a robotic engineer assistant: chatbots accessing large knowledge base enabled with NLP and speech recognition.

- Educate new generations faster on what matters most (i.e., create undergraduate oil and gas programs with an emphasis on machine learning and data analytics).

- **Challenge**: Separation among subsurface and digital disciplines. Typically, most engineering and geologic disciplines, particularly those individuals with most oilfield acumen, lack digital orientation to solve problems involving data, and process automation.

Mitigation Options

- Implement short on-the-job training programs for new employees covering key aspects of machine learning and artificial intelligence in the oil and gas industry.

- Promote cross-posting projects between technical and digital specialties.

- **Challenge**: Managing the silo mentality across multidisciplinary teams and organizations is an old challenge across industries [3]. In such conditions, individuals' and teams' lack of common goals preventing them from sharing key data and information. This mentality ultimately affects efficiency and employee morale.

Mitigation Options

- Implement collaborative project management providing visibility to ongoing work.

- Reward teamwork, favoring those who lead new ideas, those who share lessons learned, as those who protect and share knowledge.

- Provide an environment for knowledge sharing and ideas nurturing.

- **Challenge**: Managing unplanned and ad hoc work requires people to deviate from their original responsibilities, hence derailing corporate objectives. If this becomes the normal way of working, then there is a wrong functional design and/or the available (human) resources are not enough in quantity and quality.

Mitigation Options

- Implement data analytics and business processes to track ad hoc work and causes of planned vs. unplanned downtime.

Technology Challenges and Mitigations

The following highlights the most relevant technology challenges and suggested mitigating options.

- **Challenge**: Requirement to continuously manage (monitor, optimize, and control) asset performance. Oil and gas production requires a continuous understanding of the reservoir, wells, and facilities; however, these need to be achieved with a limited amount of resources, which are only able to see a fraction of the operation. This is due to a lack of sensors and technologies that allow an in-depth view of the process state.

  Mitigation Options

  - Implement data analytics and business processes to automate asset *exception-based surveillance* (EBS), focusing on critical well status.

  - Implement machine learning models to continuously search for optimum asset operating envelope, considering uncertainty.

  - Implement an automated AI-assisted complex decision-making engine with an automated routine-action recommendation.

- **Challenge**: Demonstrating a commitment to environmental concerns while providing a maximum return to shareholders.

Mitigation Options

- Enable routine HSE inspections with permanent instrumentation processing, video surveillance with digital image processing including drone and satellite images, and robotics with human interactions, focusing on critical events.

- Implement analytics processes that involve multi-objective constrained optimization objective functions looking at better fuels and profitable products.

- **Challenge**: Delivering a safer workplace with less human-intensive actions and decision-making processes. Enable operations and processes run 24/7 with very high availability, reducing human-power requirements.

Mitigation Options

- Automate production monitoring and oilfield operations

- Enable routine HSE inspections including the use of connected wearable safety devices in critical areas

- Enable autonomous and remote control of routine actions: choke, gas-lift, frequency, and so forth

Barriers to Adopting Machine Learning

These barriers may come into play before, during, and after any machine learning and artificial intelligence project execution (see Figure 8-2). These barriers, when properly identified, will be mitigated as they jeopardize sustainability and may kill the idea at any point in the project execution. They have been captured from experience and various references across the literature.

Knowledge and Skills Gaps

Understanding gaps in people skills is paramount for sustainable implementations of machine learning and artificial intelligence solutions. Multidisciplinary knowledge gaps, lack of machine learning and artificial intelligence skills, digital literacy, and anxieties over job elimination are some of the barriers to the broader adoption of these technologies in the industry.

The Knowledge Gap Between Data Scientists and Petroleum Engineers

Petroleum engineering is a science that is often forced to make decisions in the presence of little data availability; this has created a large body of knowledge for statistics, uncertainty management, and analytics. However, modern machine learning and advanced data analytics techniques in upstream are in their infancy stages.

In addition, individuals with advanced data science skills may not be well conversant with the engineering practices of the oil and gas industry.

The way to mitigate this challenge is to have people who are knowledgeable in both domains and pursue cross-domain training.

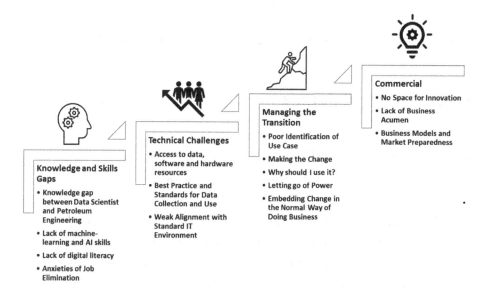

Figure 8-2. *Barriers to adopting machine learning and artificial intelligence in the oil and gas industry*

Lack of Machine Learning and Artificial Intelligence Skills

Oil and gas professionals are not conventionally trained to handle advanced machine learning and artificial intelligence technologies. As it is seen historically in many operators and services companies, a small fraction of professionals opt to get self-trained to address the application of new technologies. This is no different in machine learning and artificial intelligence, where only a few can understand the full application of such techniques.

Lack of machine-learning skills is a key obstacle, and it impedes organizations to fully exploit the value of their data.

Low availability and poor adequacy of the new workforce intensify industry sustainability goals. Young professionals may not have the required level of competencies to address the impending challenges and require long times to get trained

271

With the pressure to reduce cost and manpower, there is a need for recruiting different types of employees or a new type or generation of the workforce that makes greater use of artificial intelligence applications.

Lack of Digital Literacy

Petroleum engineers and management professionals can cope with a limited amount of digital knowledge. However, when talking about machine learning and artificial intelligence, individuals need to have a broader vocabulary, a deeper understanding of computers and knowledge of digital systems. This involves general knowledge on how to build a digital solution including all its components: defining the problem objectives, use cases, algorithm, workflow, data, and hardware and software requirements.

Scaling machine learning and artificial intelligence solutions to a wider user-based production environment is a more complex task. We believe that emerging generations will have more digital literacy and abilities to absorb more data in the digital world.

New generations will adapt to and request more data-driven applications.

Anxieties of Job Elimination

Historically, all industrial revolutions have implied some form of workforce adaptation to other skill requirements.

As identified by the skills gap, there is a genuine concern that someday machines will get super-smart and that digital transformation lead by machine learning and artificial intelligence will massively eliminate jobs. This further creates a negative response toward machine learning and artificial intelligence applications.

Technical Challenges

On the technical side, some obstacles can be easily removed with time, which includes access to machine learning resources (data, software, and hardware), lack of best practices for data use, and IT environment alignment.

Access to Data, Software, and Hardware Resources

The oil and gas industry may not have the infrastructure needed to run advanced data analytics in most of its operations. Data is typically located in the places where it is not leveraged (i.e., in the field operations where there is little knowledge about how to use data).

Cybersecurity constraints also pose an obstacle; it brings project delays until data manipulation workflows are fully understood by corporate IT personnel. Too restrictive IT policies prevent users from collaborating and sharing knowledge about machine learning applications. These policies also block users from downloading and executing open source software and programming languages, such as R and Python.

Best Practice and Standards for Data Collection and Use

Although there are numerous open source resources for machine learning and artificial intelligence applications, there is no established best practice nor standard operating procedure for gathering and using data.

Weak Alignment with Standard IT Environment

Although many software resources are open source, the lack of corporate IT environment supporting machine learning and artificial intelligence applications has typically been a concern at the time of new initiatives.

Although there is an increasing number of commercial solutions that allow an easy transition from innovation to a software development and production environment (DevOps), there is still a concern to interface with existing legacy applications and the need for sustaining the solution from an enterprise architecture point of view [4].

Managing the Transition

Managing change properly could make us winners. These transition elements include the proper identification of the use case, shaking the status-quo, providing context on why we should embrace it, letting go of power, and embedding the solutions in the existing business process.

Poor Identification of Use Case

Identifying the problem that the new solution is attempting to solve is a key step in any new application development project. Poor identification of the required use case may later destroy the efforts in machine learning application development and bring frustration to the team.

Making the Change

Being in the comfort zone makes it difficult for new applications to get into the marketplace and displace traditional applications. Shaking the status quo makes it harder for inventive applications to enter the mainstream.

Psychological Factors: Why Should I Use It?

A set of psychological factors influence any new technology adoption. These factors include rigid mindsets, risk aversion, early-adopter hesitance, lack of trust, not-invented-here attitude, jealousy, social norms, previous experiences, and organizational level factors, such as leadership and culture [5].

Studies show that explicit propensity to trust and implicit attitude toward automation did not significantly correlate [6]. Therefore, considering the analogy with machine learning and artificial intelligence, a combination of the psychological factors prevents the adoption of new applications.

Letting Go of Power

Artificial intelligence applications use data in a more democratized way, including distributing data insights accross the organization, which may jeopardize the power position of many individuals. Retaining control over specific datasets is common malpractice across industries.

Embedding Change in the Normal Way of Doing Business

One challenge is the management of change that these new applications may imply for the organization.

The management of change is of two types: (1) a new machine learning application that is changing a process that is already in place, and it is just making it more effective; and (2) a new application that creates a brand-new process. In both cases, the identified use case should clearly define the business process and governance in which the new application delivers value.

Commercial Barriers

Organizations face barriers on the commercial side because of not leaving some space for innovation, lacking business acumen, and lacking a proper business model and market preparedness to create a win-win situation between vendors and customers.

No Space for Innovation

Operators are typically in charge of operation and maintenance budgets, and they are pressured to keep them low and within the targets. On the other side, operators may not have any incentive in developing a new application that would dramatically change their way of working. This hinders any opportunity for innovation on machine learning/artificial intelligence technology application, which may ultimately add value to the bottom line.

Lack of Business Acumen

There is no value-added case sufficiently strong to support the capital expenditure on new applications. Often, engineers and operators lack financial acumen to formulate a business case.

In addition, it is difficult setting up proper success measurements or defining financial returns targets in terms of quantifiable production or revenue gains.

In consequence, organizations may not be prepared with proper budget line items, which allows them to pursue a machine learning project.

Business Models and Market Preparedness

Because machine learning and artificial intelligence applications come as new market offerings, not all software vendors have a full grasp on how the business models for their tools/solutions should look like.

This may include conversations about charging per seat vs. per asset basis, fixed fee vs. pro-bono basis, perpetual vs. leasing, transaction count vs. CPU/GPU count, volume discounts, and so forth.

These conversations may create conflicts between technology vendors and customers, which may embark on long discussions, sometimes damaging relationships.

Digital Transformation of the Oil and Gas Industry Enabled by Machine Learning and Artificial Intelligence

Machine learning and artificial intelligence may enable the digital transformation of the oil and gas industry if various contributing factors become a reality, including more talented people in machine learning/artificial intelligence professions, expanded wellsite sensing capabilities, increased use of wearable devices, implementation of better regulations, and increased use of artificial intelligence tools in government organizations.

People Skills

There are places in the oil and gas industry where machines could not easily replace the human contribution. The future petroleum engineer, released from traditional mundane tasks, would need to focus more on creative work, including innovative porous media recovery mechanisms, ground-breaking business models for the role of hydrocarbon in society, new uses of environmentally friendly materials, cost-effective facility life extension options, and so on.

The Talent Requirements for the Future

In the future, the hydrocarbon industry, enabled by digitization, analytics, and AI, will demand a more complex and innovative mindset and skillset. This will enable the ingestion of analytic insights from multiple domains. Collectively, it could promise the offer of a safer and more profitable industry, with fewer working hours and better work and personal life balance for employees.

In dealing with the digital world of today, energy professionals will carry a set of skills, including computing upgrades, system maintenance, and data manipulation that ranges from exploratory data analysis to programming of exception-based surveillance rules (see Figure 8-3).

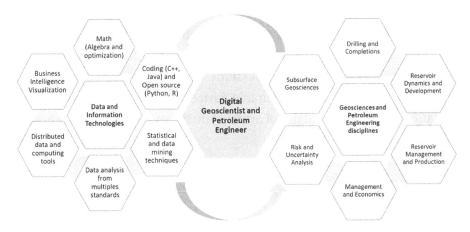

Figure 8-3. *The digital geoscientist and petroleum engineer of the future*

Contributing Factors to Further Advancement of Machine Learning

Several contributing technologies and legal frameworks may change dramatically the way machine learning and artificial intelligence are adopted in the oil and gas industry. These create additional opportunities for machine learning as well as open new paradigms in well and reservoir modeling and interpretation.

Well and Reservoir Sensing Technologies

Future increased availability of sensing technologies, including along- and beyond-the-wellbore acoustic and electromagnetic, will enable both saturation and pressure changes to be tracked in time and space. This shall open new challenging avenues for data analytics and AI.

Massive Data Collection with Social and Human Sensors

The oil and gas industry is behind other industries (e.g., finance, medical, and marketing) regarding large-scale wireless sensors [7]. These sensors may come from various sources: connected wearable devices, social networks, multi-domain collaboration, or any network of IoT (Internet-of-Things) devices.

Policymaking in AI

As machine learning and artificial intelligence take more roles in society and corporate worlds, the need to create policies by governments and organizations becomes more obvious [8]. Government policies and global cooperation agreements could accelerate (or deaccelerate) dramatically the progress and adoption of machine learning and artificial intelligence in all levels of society, including health, food, services, energy, transport and education.

Adoption of AI by Government Organizations

Governments face numerous challenges in the adoption of AI [9], including effective use of data, people skills, AI ecosystem, legacy culture, and procurement mechanisms. However, a change in the AI adoption rate by governments, such as the UAE, could imply tremendous implications in the development of machine learning/artificial intelligence for the oil and gas industry. There is a great correlation between citizen satisfaction and government adoption rates [10].

Machine Learning and Artificial Intelligence Technology Promises

There is no doubt that machine learning and artificial intelligence have progressed exponentially over the last decade [11]. This accelerated pace suggests that the associated science and techniques will continue to grow in the next decade as well [12].

AI will take normal roles in our daily lives [13], including the oil and gas industry. AI will play a key role in solving key engineering challenges of the next decade [14]; particularly, it will solve many challenges in drilling [15], production, and reservoir engineering.

Deep Learning

The performance of machine learning models increases as data availability increases. Deep neural networks have the potential to leverage more data. The following are examples of deep learning applications.

- Automating drilling operations requires the correct event classification of time-series data. An offline model based on semantic segmentation in parallel to a CNN-based inference model can classify and predict rig states in a real-time drilling analytics system with 99% accuracy [16].

- Hydraulic fracture operations continue to improve thanks to machine learning and AI-assisted complex event detection. These events include fracture stage start and end, ball seat operation identification, and categorization of periods of pump rate. A deep learning application based on CNN and U-Net architecture provides real-time automated interpretations of hydraulic fracture events [17].

- Log interpretation is frequently regarded as a time-consuming task with subjective results adding unnecessary uncertainties to reservoir studies. Automated petrophysical interpretation using multiple modeling techniques (i.e., model stacking for supervised classification) looks promising for reducing subsurface model study time-cycles and enhancing prediction quality [18].

- Identifying lithology and fluids from visual inspection of drill cuttings is required for hydrocarbon detection, well placement, subsurface navigation, and ultimately optimizing field development costs. However, such a task is usually perceived as low-added-value because it relies on subjective interpretation. The use of CNN for visual recognition has become popular in recent years in this area.

- Classifying cutting volume at a rig site in real time is achieved using deep-learning techniques [19]. Normalization and principal-component analyses (PCAs) are conducted before every video frame is fed into the classification model.

- Drill cutting image classification is difficult because of the large similarity between the image classes. A Bayesian-optimized ensemble of CNN performs superior to other known methods, particularly given the huge parameter space [20].

Reinforcement Learning

Reinforcement learning (RL) has tremendous potential where there is a large amount of data. RL is hungry for data. An application for waterflooding optimization was presented using deep reinforcement learning based on well data [21]. AI agents can learn the reservoir behavior by using the available data as pressure and phase monitoring. Trained agents can predict how the change in policies they make, affects the objective function.

Multiphysics Models

Traditional static-based and flow-based upscaling methods to generate equivalent-continuum models from the discrete-fracture model (DFM) present both low accuracy and high computational cost. Detailed fluid flow models that can run as fast as data-driven models have always been of interest and value in the oil and gas industry. A physics-based deep learning approach for the upscaling of high-resolution images of fractured media-constructed equivalent continuum models [22].

The Future of AI

Due to the programmed ability of AI to process information, there is an agreement that, in the future, AI will provide more contextual information [23] (see Table 8-1).

Table 8-1. *Various Waves of Artificial Intelligence (After [23])*

Main AI Wave	1st Wave (<2000) Describe	2nd Wave (last decade) Categorize	3rd Wave (future) Explain
	Handcrafted Knowledge	Statistical Learning	Contextual Adaptation
Main Scope	Knowledge structure defined by humans; Specifics explored by the algorithm	Statistical models for specific problem domains, trained on data	Contextual explanatory models for classes of real-world phenomena
Capabilities	Enables reasoning over narrowly defined problems No learning capability and poor handling of uncertainty	Nuanced classification and prediction capabilities No contextual capability and minimal reasoning ability	Model to drive decisions The model generates explanations of how a test character might have been created

Explainable AI

As machine learning/artificial intelligence models increase their performance, it becomes increasingly difficult to understand and explain them. The objective of explainable AI is to generate self-explained machine learning models that have equivalent superior performance (i.e., high accuracy in prediction) while enabling individuals to understand and trust these models [24].

Computational Linguists

Computational linguists are concerned with the statistical or rule-based modeling of natural language. Achieving superior human-level artificial intelligence depends on the ability to read documents directly in natural language text [25].

Considering the rise of AI-assisted conversational engines in the oil and gas industry (e.g., Nesh [26], Sandy [27], Willow [28], and others [29]), it is reasonable to believe that natural language processing will play a key role in the development of machine learning and artificial intelligence applications in the next decade.

Focused Initiatives

Sometimes taking a boiling-the-ocean approach leads to frustration and early disillusionment. Pursuing too many initiatives at the same time, and trying to prove value in all of them may lead to no results. It is then recommended to focus on a few initiatives and try to prove the value in a limited scope and short time frame. After rewards have been confirmed and communicated across stakeholders, there will be greater motivation to move forward with bigger chunks of work.

Data as an Asset

Without large databases, it is not easy to build strong machine learning applications. Therefore, this will be limited to a few companies unless industries decide to share publicly meaningful sets of data (e.g., Norwegian Volve field data set) to collaborate and build robust applications. Not even giant service companies nor operators have the full data set to understand the complexity of oilfield operations. This may have implications for either cost reduction of the supply chain or appearance of new lines of business in the oil patch that make money from advertising, suppliers, and companies.

Getting More Ideas Adopted

Machine learning and artificial intelligence applications are emerging as a new class of engineering or scientific method. A lot of what is done today is experimental steps to search for the perfect algorithm. Oil and gas companies have merely started to spend significant effort in developing a blueprint of data-driven applications.

There are many initiatives in the ideas and proof-of-concept stage; however, only a few ideas will be piloted on a small scale, and a small fraction will be rolled out as mainstream products.

In the meantime, the oil and gas industry needs to exercise a lot of trial and error. Without trying, the oil and gas industry will not achieve what other industries have achieved. For this, a growth mindset is needed.

Maybe some ideas are not ready for prime time, but trial and error are required.

Summary

The oil and gas industry will continue to play a key role in our society over the next decades, and its sustainability depends on our ability to address and mitigate important processes, people, and technology challenges.

The oil and gas industry's "process" challenges include managing downtime, interconnecting its business process, and optimizing the hydrocarbon value chain to provide a sustainable industry and secure energy source for society.

People challenges include low availability and poor adequacy of the workforce to solve ongoing problems, exacerbated by separation among subsurface and digital disciplines, managing the silo mentality across multidisciplinary teams and organizations, and prioritizing unplanned and ad hoc work.

Technology-related challenges include requirements to continuously manage (monitor, optimize and control) asset performance, demonstrating a commitment to environmental concerns while providing a maximum return to shareholders and delivering a safer workplace with less human-intensive actions and decision-making processes.

Machine learning and artificial intelligence technology applications are still in the infancy stage in the oil and gas industry; therefore, tremendous opportunities exist for young professionals to perform and find ways to grow their careers. Major barriers exist in terms of knowledge skills and gaps, technical challenges including lack of standards and weak alignment with traditional IT, and changes including clear use case definitions and embedding the change into the existing processes.

On the commercial side, machine learning and artificial intelligence technologies still lack an industry framework, where there is a balance between innovation and sustainable business models. This is not only seen in the oil and gas industry but throughout many industries.

The evolution of key digital technologies will enable a better road for machine learning and artificial intelligence technologies in the oil and gas industry, including cost effective well site sensors and massive data capture through wearable devices.

The continuous progress of machine learning and artificial intelligence core technologies, including deep learning and explainable AI, is a key factor for enhancing usability in many industries, including the oil and gas industry, which will enable more transformative and value-added processes.

Policymaking toward machine learning and artificial intelligence use, in addition to increased adoption through large government organizations, will enhance the use of machine learning and artificial intelligence in all sectors. Therefore, it will provide oxygen to the oil and gas industry for navigating through the next decades.

References

[1] "Outlook for Energy," US Energy Information Administration, September 24, 2019. [Online]. Available: `https://www.eia.gov/outlooks/ieo/`.

[2] G. Zang, L. Neuhofer, D. Zabel, P. Tippel, C. I. Pantazescu, V. Krcmarik, L. Krenn, and B. Hachmöller, "Smart and Automated Workover Candidate Selection," in SPE Intelligent Energy International Conference and Exhibition, Aberdeen, Scotland, UK, 2016.

[3] B. Gleeson, "The Silo Mentality: How to Break Down the Barriers," *Forbes*, October 2, 2013. [Online]. Available: `https://www.forbes.com`.

[4] C. Procaccini, "Application of Data Analytics Technologies to Improve Asset Operations and Maintenance: Digital Landscaping Study of the Oil and Gas Sector," March 9, 2018. [Online]. Available: `https://www.theogtc.com/media/2380/`.

[5] R. Roberts and R. Flin, "Unlocking the Potential: Understanding the Psychological Factors That Influence Technology Adoption in the Upstream Oil and Gas Industry," *SPE Journal*, vol. 25, no. 1, pp. 515–528, February 1, 2020.

[6] Stephanie M. Merritt, Heather Heimbaugh, Jennifer and Jennifer LaChapell, Deborah Lee, "I Trust It, but I Don't Know Why: Effects of Implicit Attitudes Toward Automation on Trust in an Automated System," Human Factors: The Journal of the Human Factors and Ergonomics Society, vol. 55, no. 3, pp. 520-534, 2013.

[7] T. Zhu, S. Xiao, Q. Zhang, and Y. Gu, "Emergent Technologies in Big Data Sensing: A Survey," International Journal of Distributed Sensor Networks, vol. 11, no. 10, 2015.

[8] B. Perry and R. Uuk, "AI Governance and the Policymaking Process: Key," Big data and Cognitive Computing, vol. 3, no. 26, p. 17, 2019.

[9] J. Torres Santeli and S. Gerdon, "World Economic Forum," 16 August 2019. [Online]. Available: `https://www.weforum.org/agenda/2019/08/artificial-intelligence-government-public-sector/`.

[10] M. Carraszo, S. Mills, A. Whybrew, and A. Jura, "The Citizen's Perspective on the Use of AI in Government," 1 March 2019. [Online]. Available: https://www.bcg.com/publications/2019/citizen-perspective-use-artificial-intelligence-government-digital-benchmarking.aspx.

[11] J. Markoff, "Divining the future: Special Report: The Rapid Advance of Artificial Intelligence," October 14, 2013. [Online]. Available: https://www.nytimes.com/2013/10/15/technology/the-rapid-advance-of-artificial-intelligence.html.

[12] S. Hansen, "What to Expect with The Future of AI Technology," April 5, 2019. [Online]. Available: https://hackernoon.com/what-to-expect-with-the-future-of-ai-technology-782aec311a54.

[13] G. Templeton, "25 Examples of AI That Will Seem Normal in 2027: From Cooking to Dating to Art," May 15, 2017. [Online]. Available: https://www.inverse.com/article/31340-ai-machine-learning-list-change-life-decade.

[14] "10 Major Engineering Challenges of the Next Decade," *Elsevier*, [Online]. Available: https://www.elsevier.com/rd-solutions/industry-insights/other/10-major-engineering-challenges-of-the-next-decade.

[15] N. A. Nunoo, "Guest Editorial: How Artificial Intelligence Will Benefit Drilling?" *JPT*, May 1, 2018.

[16] Y. Ben, W. Han, C. James, and D. Cao, "Building a General and Sustainable Machine Learning Solution in a Real-Time Drilling System," Society of Petroleum Engineers, February 25, 2020.

[17] Y. Shen, D. Cao, K. Ruddy, and L. F. Teixeira de Moraes, "Near Real-Time Hydraulic Fracturing Event Recognition Using Deep Learning Methods," *SPE Drilling & Completion*, vol. 35, no. 1, 2020.

[18] D. Yuan, Y. Li, W. Zhang, H. Wu, and W. Wang, "Automatic Reservoir Interpretation from Conventional Well Logs Using Stacking Machine Learning Technique," International Petroleum Technology Conference, January 2020.

[19] X. Du, Y. Jin, X. Wu, Y. Liu, X. Wu, O. Awan, and Z. Han, "Classifying Cutting Volume at Shale Shakers in Real-Time Via Video Streaming Using Deep-Learning Techniques.," Society of Petroleum Engineers. February 1, 2020.

[20] M. Kathrada and B. J. Adillah, "Visual Recognition of Drill Cuttings Lithologies Using Convolutional Neural Networks to Aid Reservoir Characterization," *Society of Petroleum Engineers Journal*, September 2019.

[21] R. Miftakhov, A. Al-Qasim, and I. Efremov, " Deep Reinforcement Learning: Reservoir Optimization from Pixels," International Petroleum Technology Conference, January 2020.

[22] X. He, R. Santoso, and H. Hoteit, "Application of Machine-Learning to Construct Equivalent Continuum Models from High-Resolution Discrete-Fracture Models," International Petroleum Technology Conference, January 2020.

[23] J. Launchbury, "A DARPA Perspective on Artificial Intelligence," DARPA, 15 February 2017. [Online]. Available: https://www.darpa.mil/about-us/darpa-perspective-on-ai.

[24] M. Turek, "Artificial Intelligence Colloquium: Explainable AI," [Online]. Available: https://www.darpa.mil/program/explainable-artificial-intelligence.

[25] J. McCarthy, "What AI Needs from Computational Linguistics," Stanford University, [Online]. Available: http://jmc.stanford.edu/index.html.

[26] hellonesh.io, "Hello Nesh," [Online]. Available: https://hellonesh.io/.

[27] Belmont Technology Inc., Belmont Technology Inc., [Online]. Available: https://www.b15y.io/.

[28] IBM, "IBM Woodside Willow," August 12, 2018. [Online]. Available: https://youtu.be/BocVnDrmtZo.

[29] Baker Hughes, "AI Capabilities for Oil and Gas," June 28, 2018. [Online]. Available: https://youtu.be/X9k-gLwDJ9g.

[30] World Economic Forum, "White Paper: Digital Transformation Initiative: Oil and Gas Industry," 2017.

Index

© Yogendra Narayan Pandey, Ayush Rastogi, Sribharath Kainkaryam,
Srimoyee Bhattacharya, and Luigi Saputelli 2020
Y. N. Pandey et al., *Machine Learning in the Oil and Gas Industry*,
https://doi.org/10.1007/978-1-4842-6094-4

Printed in the United States
By Bookmasters